An Introduction to Logic Circuit Testing

An Introduction to Logic Circuit Testing
Parag K. Lala

ISBN: 978-3-031-79784-2   paperback

ISBN: 978-3-031-79785-9   ebook

DOI: 10.1007/978-3-031-79785-9

A Publication in the Springer series

SYNTHESIS LECTURES ON DIGITAL CIRCUITS AND SYSTEMS # 17
Lecture #17

Series Editor: Mitchell Thornton, Southern Methodist University

Series ISSN
ISSN 1932-3166   print
ISSN 1932-3174   electronic

An Introduction to
Logic Circuit Testing

Parag K. Lala
Texas A&M University–Texarkana

SYNTHESIS LECTURES ON DIGITAL CIRCUITS AND SYSTEMS # 17

ABSTRACT

An Introduction to Logic Circuit Testing provides a detailed coverage of techniques for test generation and testable design of digital electronic circuits/systems. The material covered in the book should be sufficient for a course, or part of a course, in digital circuit testing for senior-level undergraduate and first-year graduate students in Electrical Engineering and Computer Science. The book will also be a valuable resource for engineers working in the industry. This book has four chapters. Chapter 1 deals with various types of faults that may occur in very large scale integration (VLSI)-based digital circuits. Chapter 2 introduces the major concepts of all test generation techniques such as redundancy, fault coverage, sensitization, and backtracking. Chapter 3 introduces the key concepts of testability, followed by some ad hoc design-for-testability rules that can be used to enhance testability of combinational circuits. Chapter 4 deals with test generation and response evaluation techniques used in BIST (built-in self-test) schemes for VLSI chips.

KEYWORDS

digital circuits, logic circuit testing, VLSI, fault detection, design-for-testability, response evaluation techniques, BIST, D-Algorithm, PODEM, FAN, LFSR

Dedication

To my wife Meena

What lies behind us and lies before us are tiny matters compared to what lies within us.
Ralph Waldo Emerson

Preface

This book provides a detailed coverage of techniques for test generation and testable design of digital electronic circuits/systems. The material covered in the book should be sufficient for a course or part of a course in digital circuit testing for senior-level undergraduate and first-year graduate students in Electrical Engineering and Computer Science.

This book has four chapters. Chapter 1 deals with various types of faults that may occur in very large scale integration (VLSI)-based digital circuits. The modeling of faults at the gate level and at the transistor level is considered. It discusses the fundamental concepts of fault detection and also introduces the concepts of controllability, observability, and fault equivalency. The chapter finishes with a brief discussion of temporary faults.

Chapter 2 introduces the major concepts of all test generation techniques such as redundancy, fault coverage, sensitization, and backtracking. It examines in detail various techniques available for fault detection in combinational logic circuits such as D-algorithm, PODEM, and FAN. This chapter also discusses test generation for sequential circuits. It covers the state table verification approach for fault detection in sequential circuits as well as a technique that utilizes both structure and function (state table) of sequential circuits.

Chapter 3 introduces the key concepts of testability, followed by some ad hoc design-for-testability rules that can be used to enhance testability of combinational circuits. This chapter also covers in detail major design methods for enhancing testability of sequential circuits implemented on VLSI-based digital systems.

Chapter 4 deals with test generation and response evaluation techniques used in built-in self-test (BIST) schemes for VLSI chips. Because linear feedback shift register (LFSR)-based techniques are used in practice to generate test patterns and evaluate output responses in BIST, such techniques are thoroughly discussed. In addition, some popular BIST architectures are examined.

Contents

CHAPTER 1

Introduction

1.1   FAULTS IN LOGIC CIRCUITS

A *failure* is said to have occurred in a logic circuit or system if it deviates from its specified behavior [1]. A *fault*, on the other hand, refers to a physical defect in a circuit. For example, a short between two signal lines in the circuit or a break in a signal line is a physical defect. An *error* is usually the manifestation of a fault in the circuit; thus a fault may change the value of a signal in a circuit from 0 (correct) to 1 (erroneous) or vice versa. However, a fault does not always cause an error; in that case, the fault is considered to be *latent*.

A fault is characterized by its *nature, value, extent,* and *duration* [2]. The nature of a fault can be classified as *logical* or *nonlogical*. A logical fault causes the logic value at a point in a circuit to become opposite to the specified value. Nonlogical faults include the rest of the faults such as the malfunction of the clock signal, power failure, etc. The value of a logical fault at a point in the circuit indicates whether the fault creates fixed or varying erroneous logical values. The extent of a fault specifies whether the effect of the fault is localized or distributed. A local fault affects only a single variable, whereas a distributed fault affects more than one. A logical fault, for example, is a local fault, whereas the malfunction of the clock is a distributed fault. The duration of a fault refers to whether the fault is *permanent* or *temporary*.

1.1.1   Stuck-At Fault

The most common model used for logical faults is the *single stuck-at fault*. It assumes that a fault in a logic gate results in one of its inputs or the output is fixed at either a logic 0 (*stuck-at-0*) or at logic 1 (*stuck-at-1*). Stuck-at-0 and stuck-at-1 faults are often abbreviated to *s-a-0* and *s-a-1*, respectively.

Let us assume that in Figure 1.1 the *A* input of the NAND gate is s-a-1. The NAND gate perceives the *A* input as a logic 1 irrespective of the logic value placed on the input. For example, the output of the NAND gate is 0 for the input pattern *A*=0 and *B*=1, when input *A* is s-a-1 in. In the absence of the fault, the output will be 1. Thus, *AB*=01 can be considered as the *test* for the *A* input s-a-1, since there is a difference between the output of the fault-free and faulty gate.

The single stuck-at fault model is often referred to as the *classical fault model* and offers a good representation for the most common types of defects [e.g., shorts and opens in complementary

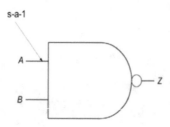

FIGURE 1.1: Two-input NAND gate.

metal oxide semiconductor (CMOS) technology]. Figure 1.2 illustrates the CMOS realization of the two-input NAND:

The number 1 in the figure indicates an open, whereas the numbers 2 and 3 identify the short between the output node and the ground and the short between the output node and the V_{DD}, respectively. A short in a CMOS results if not enough metal is removed by the photolithography, whereas over-removal of metal results in an open circuit [3]. Fault 1 in Figure 1.2 will disconnect input A from the gate of transistors $T1$ and $T3$. It has been shown that in such a situation one transistor may conduct and the other remain nonconducting [4]. Thus, the fault can be represented by a stuck at value of A; if A is s-a-0, $T1$ will be ON and $T3$ OFF, and if A is s-a-1, $T1$ will be OFF and $T3$ ON. Fault 2 forces the output node to be shorted to V_{DD}, that is, the fault can be considered as an s-a-1 fault. Similarly, fault 3 forces the output node to be s-a-0.

The stuck-at model is also used to represent multiple faults in circuits. In a multiple stuck-at fault, it is assumed that more than one signal line in the circuit are stuck at logic 1 or logic 0; in other

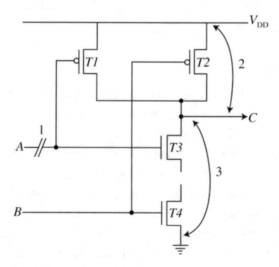

FIGURE 1.2: Two-input NAND gate in CMOS gate.

words, a group of stuck-at faults exist in the circuit at the same time. A variation of the multiple fault is the *unidirectional fault*. A multiple fault is unidirectional if all of its constituent faults are either s-a-0 or s-a-1 but not both simultaneously. The stuck-at model has gained wide acceptance in the past mainly because of its relative success with small scale integration. However, it is not very effective in accounting for all faults in present day very large scale integrated (VLSI), circuits which mainly uses CMOS technology. Faults in CMOS circuits do not necessarily produce logical faults that can be described as stuck-at faults [5, 6, 7]. For example, in Figure 1.2, faults 3 and 4 create stuck-on transistors faults. As a further example, we consider Figure 1.3, which represents CMOS implementation of the Boolean function:

$$Z = \overline{(A+B)(C+D) \cdot EF}.$$

Two possible shorts numbered 1 and 2 and two possible opens numbered 3 and 4 are indicated in the diagram. Short number 1 can be modeled by s-a-1 of input E; open number 3 can be modeled by s-a-0 of input E, input F, or both. On the other hand, short number 2 and open number

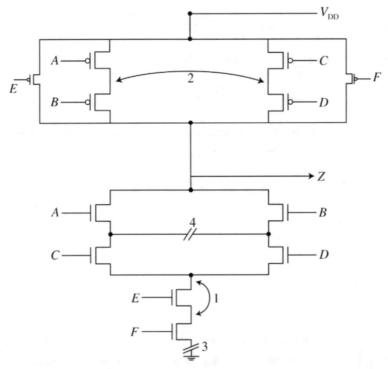

FIGURE 1.3: CMOS implementation of $Z = \overline{(A+B)(C+D) \cdot EF}$.

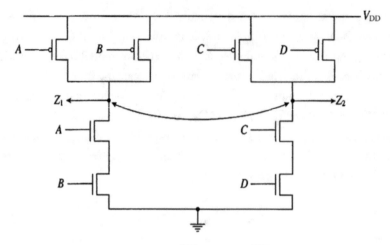

FIGURE 1.4: CMOS implementation of $Z_1 = \overline{AB}$ and $Z_2 = \overline{CD}$.

4 cannot be modeled by any stuck-at fault because they involve a modification of the network function. For example, in the presence of short number 2, the network function will change to:

$$Z = \overline{(A+C)(B+D) \cdot EF},$$

and open number 4 will change the function to:

$$Z = \overline{(AC) + (BD) \cdot EF}.$$

For this reason, a perfect short between the output of the two gates (Figure 1.4) cannot be modeled by a stuck-at fault. Without a short, the outputs of gates Z_1 and Z_2 are:

$$Z_1 = \overline{AB} \quad \text{and} \quad Z_2 = \overline{CD},$$

whereas with the short,

$$Z_1 = Z_2 = \overline{AB} + \overline{CD}.$$

1.1.2   Bridging Faults

Bridging faults form an important class of permanent faults that cannot be modeled as stuck-at faults. A bridging fault is said to have occurred when two or more signal lines in a circuit are accidentally connected together. Earlier study of bridging faults concentrated only on the shorting of signal lines in gate-level circuits. It was shown that the shorting of lines resulted in *wired logic* at the connection.

Bridging faults at the gate level has been classified into two types: *input bridging* and *feedback bridging*. An input bridging fault corresponds to the shorting of a certain number of primary input lines. A feedback bridging fault results if there is a short between an output and input line. A feedback bridging fault may cause a circuit to oscillate, or it may convert it into a sequential circuit.

Bridging faults in a transistor-level circuit may occur between the terminals of a transistor or between two or more signal lines. Figure 1.5 shows the CMOS logic realization of the Boolean function:

$$Z_1 = Z_2 = \overline{AB} + \overline{CD}$$

A short between two lines, as indicated by the dotted line in the diagram will change the function of the circuit.

The effect of bridging among the terminals of transistors is technology-dependent. For example, in CMOS circuits, such faults manifest as either stuck-at or stuck-open faults, depending on the physical location and the value of the bridging resistance.

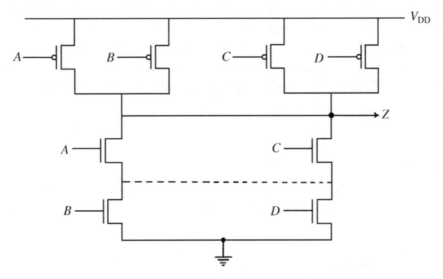

FIGURE 1.5: CMOS implementation of $Z(A, B, C, D) = \overline{AB} + \overline{CD}$

1.1.3   Delay Faults

As mentioned previously, not all manufacturing defects in VLSI circuits can be represented by the stuck-at fault model. The size of a defect determines whether the defect will affect the logic function of a circuit. Smaller defects, which are likely to cause partial open or short in a circuit, have a higher probability of occurrence due to the statistical variations in the manufacturing process [8]. These defects result in the failure of a circuit to meet its timing specifications without any alteration of the logic function of the circuit. A small defect may delay the transition of a signal on a line either from 0 to 1, or vice versa. This type of malfunction is modeled by a *delay fault*.

Two types of delay faults have been proposed in literature: *gate delay fault* and *path delay fault*. Gate delay faults have been used to model defects that cause the actual propagation delay of a faulty gate to exceed its specified worst case value. For example, if the specified worst case propagation delay of a gate is x units and the actual delay is $x+\Delta x$ units, then the gate is said to have a delay fault of size Δx. The main deficiency of the gate delay fault model is that it can only be used to model isolated defects, not distributed defects, for example, several small delay defects. The path delay fault model can be used to model isolated as well as distributed defects. In this model, a fault is assumed to have occurred if the propagation delay along a path in the circuit under test exceeds the specified limit.

1.2   BREAKS AND TRANSISTORS STUCK-OPEN AND STUCK-ON OR STUCK-OPEN FAULTS IN CMOS

As discussed previously, not all defects in CMOS VLSI can be represented by using the stuck-at fault model. It has been shown that breaks and transistor stuck-ons are two other types of defects that, like bridging, may remain undetected if testing is performed based on the stuck-at fault assumption. These defects have been found to constitute a significant percentage of defects occurring in CMOS circuits [2]. In the following two subsections, we discuss the effects of these defects on CMOS circuits.

1.2.1   Breaks

Breaks or opens in CMOS circuits are caused either by missing conducting material or extra insulating material. Breaks can be either of the following two types [3]:

1.   Intragate breaks;
2.   Signal line breaks.

An intragate break occurs internal to a gate. Such a break can disconnect the source, the drain, or the gate from a transistor, identified by b_1, b_2, and b_3, respectively, in Figure 1.6. The presence of b_3, will have no logical effect on the operation of a circuit, but it will increase the propagation delay; that is, the break will result in a delay fault. Similarly, the break at b_1 will also produce a delay fault without changing the function of the circuit. However, the break at b_2 will make the p-transistor nonconducting; that is, the transistor can be assumed to be *stuck-open*.

An intragate break can also disconnect the p-network, the n-network, or both networks (b_4, b_5, and b_6 in Figure 1.6) from the circuit. The presence of b_4 or b_5 will have the same effect as the output node getting stuck-at-0 or stuck-at-1, respectively. In the presence of b_6, the output voltage may have an intermittent stuck-at-1 or stuck-at-0 value; thus, if the output node simultaneously drives a p-transistor and an n-transistor, then one of the transistors will be ON for some unpredictable period of time. Signal line breaks can force the gates of transistors in static CMOS circuits to float.

As shown in Figure 1.6, such a break can make the gate of only a p-transistor and an n-transistor to float. It is also possible, depending on the position of a break, that the gates of both transistors may float, in which case one transistor may conduct and the other remain in a

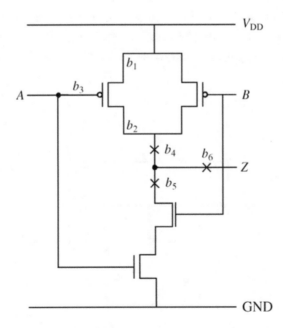

FIGURE 1.6: Two-input CMOS NAND gate showing occurrence of breaks.

nonconducting state [9]. In general, this type of break can be modeled as a stuck-at fault. On the other hand, if two transistors with floating gates are permanently conducting, one of them can be considered as stuck-on. If a transistor with a floating gate remains in a nonconducting state due to a signal line break, the circuit will behave in a similar fashion as it does in the presence of the intragate break b_2.

1.2.2   Stuck-On and Stuck-Open Faults

A *stuck-on* transistor fault implies the permanent closing of the path between the source and the drain of the transistor. Although the stuck-on transistor, in practice, behaves in a similar way as a *stuck-closed* transistor, there is a subtle difference. A stuck-on transistor has the same drain-source resistance as the on resistance of a fault-free transistor, whereas a stuck-closed transistor exhibits a drain-source resistance that is significantly lower than the normal on-resistance. In other words, in the case of stuck-closed transistor, the short between the drain and the source is almost perfect, and this is not true for a stuck-on transistor. A transistor stuck-on (stuck-closed) fault may be modeled as a bridging fault from the source to the drain of a transistor.

A *stuck-open* transistor implies the permanent opening of the connection between the source and the drain of a transistor. The drain-source resistance of a stuck-open transistor is significantly higher than the off-resistance of a nonfaulty transistor. If the drain-source resistance of a faulty transistor is approximately equal to that of a fault-free transistor, then the transistor is considered to be *stuck-off*. For all practical purposes, transistor stuck-off and stuck-open faults are functionally equivalent.

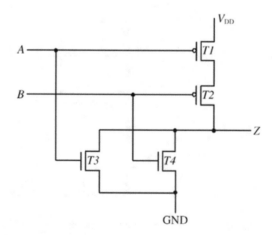

FIGURE 1.7: A two-input CMOS NOR gate.

TABLE 1.1: Truth table of two-input CMOS NOR gate with and without stuck-open fault

A	B	Z	Z (As-op)	Z (Bs-op)	Z (V$_{DD}$s-op)
0	0	1	1	1	Z_t
0	1	0	0	Z_t	0
1	0	0	Z_t	0	0
1	1	0	0	0	0

A stuck-open transistor fault like a feedback bridging fault can turn a combinational circuit into a sequential circuit [10]. Figure 1.7 shows a two-input CMOS NOR gate. A stuck-open fault causes the output to be connected neither to GND nor to V_{DD}. If, for example, transistor $T2$ is open-circuited, then for input $AB=00$, the pull-up circuit will not be active and there will be no change in the output voltage. In fact, the output retains its previous logic state; however, the length of time the state is retained is determined by the leakage current at the output node.

Table 1.1 shows the truth table for the two-input CMOS NOR gate. The fault-free output is shown in column Z; the three columns to the right represent the outputs in presence of the three stuck-open (s-op) faults. The first, *As-op*, is caused by any input, drain, or source missing connection to the pull-down FET $T3$. The second, Bs-op, is caused by any input, drain, or source missing connection to the pull-down FET $T4$. The third, $V_{DD}s$-op, is caused by an open anywhere in the series, p-channel pull-up connection to VDD. The symbol Z_t is used to indicate that the output state retains the previous logic value.

1.3   BASIC CONCEPTS OF FAULT DETECTION

Fault detection in a logic circuit is carried out by applying a sequence of tests and observing the resulting outputs. A *test* is an input combination that specifies the expected response that a fault-free circuit should produce. If the observed response is different from the expected response, a fault is

FIGURE 1.8: A NAND gate with a stuck-at-1 fault.

TABLE 1.2: Output response of the NAND gate

IINPUT		OUTPUT	
a	b	c (FAULT-FREE)	c (FAULT-PRESENT)
0	0	1	1
0	1	1	0
1	0	1	1
1	1	0	0

present in the circuit. The aim of testing at the gate level is to verify that each logic gate in the circuit is functioning properly and the interconnections are good. Henceforth, we will deal with stuck-at faults only unless mentioned otherwise. If only a single stuck-at fault is assumed to be present in the circuit under test, then the problem is to construct a test set that will detect the fault by utilizing only the inputs and the outputs of the circuit.

As indicated above, a test detects a fault in a circuit if and only if the output produced by the circuit in the presence of the fault is different from the observed output when the fault is not present. To illustrate, let us assume that input a of the NAND gate shown in Figure 1.8 is stuck-at-1. The output responses of the gate to all input combinations for both fault-free and fault-present conditions are shown in Table 1.2.

It can be seen in Table 1.2 that only for input combination $ab=0$, the output is different in the presence of the fault a s-a-1 and when the gate is fault-free.

In order to detect a fault in a circuit, the fault must first be *excited*; that is, a certain input combination must be applied to the circuit so that the logic value appearing at the fault location is opposite to the fault value. Next, the fault must be *sensitized*; that is, the effect of the fault is propa-

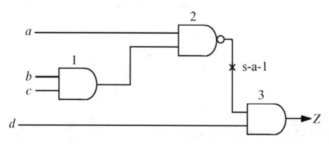

FIGURE 1.9: Circuit with a single s-a-1 fault.

gated through the circuit to an observable output. For example, in Figure 1.9, the input combination abc=111 must be applied for the excitation of the fault, and d=1 for sensitizing the fault to output Z. Thus, the test for the s-a-1 fault is $abcd$=1111. This input combination is also a test for other faults (e.g., gate 1 s-a-0, gate 3 s-a-1, and input a s-a-0, etc.).

1.3.1   Controllability and Observability

As indicated above, in order to generate a test a for a stuck-at fault on a signal line, it must first be forced to a value that is opposite to the stuck-at value on the line. This ability to apply input patterns to the primary inputs of a circuit to set up appropriate logic value at desired locations of a circuit is known as *controllability*. For example, in the presence of a stuck-at-0 fault, the location of the fault must be set to logic 1 via the primary inputs; this is known as *1-controllability*. Similarly, for a stuck-at-1 fault, the location of the fault must be set to logic 0 to excite the fault; this is known as *0-controllability*.

The sensitization part of the test generation process requires application of appropriate input values at the primary inputs so the effect of the fault is observable at the primary outputs. For example, in Figure 1.9, the effect of the stuck-at-1 fault can observed at output Z if input d is set at 1; if d is set to 0, the output will be 0 and the effect of the fault will be masked (i.e., the fault will not be detected). The ability to observe the response of a fault on an internal node via the primary outputs of a circuit is denoted as *observability*.

1.3.2   Undetectable Faults

A fault is considered to be undetectable if it is not possible to activate the fault or to sensitize its effect to primary outputs. In other words, a test for detecting the fault does not exist. To illustrate, let us consider the α s-a-0 fault shown in Figure 1.10. It is not possible to set the node α to logic 1. Therefore, the fault cannot be excited and thus undetectable. The fault β s-a-0 can be excited by making ab=10, but no sensitized path is available for propagating the effect of the fault to the

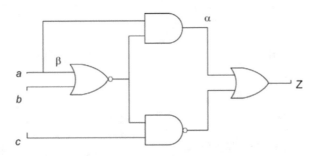

FIGURE 1.10: Circuit with a stuck-at-0 faults.

output; hence, the fault is undetectable. A combinational circuit is denoted as *redundant* if it has an undetectable fault.

A test set for a circuit is derived based on the assumption that only a single fault is present in the circuit when the tests are applied. Thus, the simultaneous presence of an undetectable fault and a detectable fault violates this assumption. Furthermore, the presence of an undetectable fault may prevent the detection of a detectable fault.

1.3.3   Equivalent Faults

A test, in general, can detect more than one fault in a circuit, and many tests in a set detect the same faults. In other words, the subsets of faults detected by each test from a test set are not disjoint. Thus, a major objective in test generation is to reduce the total number of faults to be considered by grouping equivalent faults in subsets. It is then sufficient only to test one fault from each equivalent set to cover all faults in the set, thus avoiding redundancy in the test generation process.

In an m-input gate, there can be $2(m+1)$ stuck-at faults. Thus, the total number of single stuck-at faults in a two-input NOR gate shown in Figure 1.11a is 6 (=2×3), e.g,. a s-a-0, b s-a-0, a s-a-1, b s-a-1, c s-a-0 and c s-a-1. However, a stuck-at fault on an input may be indistinguishable from a stuck-at fault at the output. For example, in a NOR gate (Figure 1.11a), any input s-a-1 fault is indistinguishable from the output s-a-0; similarly, in a NAND gate (Figure 1.11b), an input s-a-0 fault is indistinguishable from the output s-a-1.

Two faults are considered to be *equivalent* if every test for one fault also detects the other. In the two-input NOR gate shown in Figure 1.11, a stuck-at-1 fault on one of the inputs a or b is equivalent to output c stuck-at-0, thus all three faults belong to the same equivalence set. A test for any of these three faults will also detect the presence of the other two. The equivalence sets for the NOR gate are:

$$\{a \text{ s-a-1}, b \text{ s-a-1}, c \text{ s-a-0}\},$$
$$\{a \text{ s-a-0}, c \text{ s-a-1}\},$$
$$\{b \text{ s-a-0}, c \text{ s-a-1}\},$$

(a)   (b)

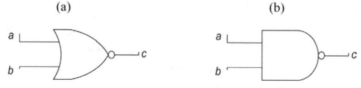

FIGURE 1.11: (a) NOR gate. (b) NAND gate.

and the equivalence sets for the NAND gate are:

$$\{a \text{ s-a-0}, b \text{ s-a-0}, c \text{ s-a-1}\},$$
$$\{a \text{ s-a-1}, c \text{ s-a-0}\},$$
$$\{b \text{ s-a-1}, c \text{ s-a-0}\}.$$

Because there are three equivalence fault sets for both NOR and NAND gates, it is sufficient to derive tests for three faults only in each case, i.e., one fault from each set. In general, an m-input gate can have a total of $(m+2)$ logically distinct faults; however, only $m+1$ equivalent sets of faults need to be considered.

1.3.4 Temporary Faults

As stated earlier, an error is a manifestation of a fault. A temporary fault can result in an *intermittent* or a *transient* error. Transient errors are the major source of failures in VLSI chips. They are nonrecurring and are not repairable because there is no physical damage to the hardware. Very deep submicron technology has enabled the packing of millions of transistors on a VLSI chip by reducing the transistor dimensions. However, the reduction of transistor sizes also reduces their noise margins. As a result, they become more vulnerable to noise, cross-talk, etc., which in turn result in transient errors. In addition, small transistors are affected by terrestrial radiation and suffer temporary malfunction, thereby increasing the rate of transient errors.

Intermittent faults are recurring faults that reappear on a regular basis. Such faults can occur due to loose connections, partially defective components, or poor designs. Intermittent faults occurring due to deteriorating or aging components may eventually become permanent. Some intermittent faults also occur due to environmental conditions such as temperature, humidity, vibration, etc. The likelihood of such intermittent faults depends on how well the system is protected from its physical environment through shielding, filtering, cooling, etc. An intermittent fault in a circuit causes a malfunction of the circuit only if it is active; if it is inactive, the circuit operates correctly. A circuit is said to be in a fault active state if a fault present in the circuit is active, and it is said to be in the fault-not-active state if a fault is present but inactive [11]. Because intermittent faults are random, they can be modeled only by using probabilistic methods.

REFERENCES

[1] Anderson, T., and P. Lee, *Fault-Tolerance: Principles and Practice*, Prentice-Hall International (1981).

[2] Avizienis, A., "Fault-tolerant systems," *IEEE Trans. Comput.*, 1304–11 (December 1976).

[3] Shoji, M., *CMOS Digital Circuit Technology*, Prentice-Hall (1988).

[4] Maly, W., P. Nag, and P. Nigh, "Testing oriented analysis of CMOS ICs with opens," *Proc. Intl. Conf. CAD*, 344–7 (1988). doi:10.1109/ICCAD.1988.122525

[5] Ferguson, I. and J. Shen, "A CMOS fault extractor for inductive fault analysis," *IEEE Trans. CAD*, 1181–94 (November 1988). doi:10.1109/43.9188

[6] David, M. W., "An optimized delay testing technique for LSSD-based VLSI logic circuits," *IEEE VLSI Test Symp.*, 239–46 (1991).

[7] Wadsack, R. L., "Fault modelling and logic simulation of CMOS and MOS integrated circuits," *Bell Syst. Technol. Jour.*, 1149–75 (May–June 1978).

[8] Ferguson. J., M. Taylor, and T. Lamabee, "Testing for parametric faults in static CMOS circuits," *Proc. Intl. Test Conf.*, 436–42 (1990). doi:10.1109/TEST.1990.114052

[9] Maly, W., "Realistic fault modeling for VUI testing," *Proc. 24th ACMI IEEE Design Auzomatlon Conf.*, 173–80 (1987).

[10] Ferguson, J., and J. Shen, "Extraction and simulation of realistic CMOS faults using inductive fault analysis," *Proc. Intl. Test Conf.*, 475–84 (1988). doi:10.1109/TEST.1988.207759

[11] Malaiya, Y. K., and S. Y. H. Su, "A survey of methods for intermittent fault analysis," *Proc. Nut Comput Conf.*, 577–84 (1979).

CHAPTER 2

Fault Detection in Logic Circuits

The aim of testing at the gate level is to verify that each logic gate in the circuit is functioning properly and the interconnections are good. If only a single stuck-at fault is assumed to be present in the circuit under test, then the problem is to construct a test set that will detect the fault by utilizing only the inputs and the outputs of the circuit.

One of the main objectives in testing is to minimize the number of test patterns. If the function of a circuit in the presence of a fault is different from its normal function (i.e., the circuit is *nonredundant*), then an n-input combinational circuit can be completely tested by applying all 2^n combinations to it; however, 2^n increases very rapidly as n increases. For a sequential circuit with n inputs and m flip-flops, the total number of input combinations necessary to exhaustively test the circuit is $2^n \times 2^m = 2^{m+n}$. If, for example, $n=20$ and $m=40$, there would be 2^{60} tests. At a rate of 10,000 tests per second, the total test time for the circuit would be about 3.65 million years! Fortunately, a complete truth table exercise of the logic circuit is not necessary–only the input combinations that detect most of the faults in the circuit are required.

The efficiency of a test set is measured by a figure of merit called *fault coverage*. The term *fault coverage* refers to the percentage of the possible single stuck-at faults that a test set will detect. The computation time needed to generate tests for combinational circuits is proportional to the square of the number of gates in the circuit. For example, the test generation time for a 100,000-gate circuit is 100 times that for a 10,000-gate circuit. The task is even more complicated for sequential circuits because the number of internal states is an exponential function of the number of memory elements. Thus, sequential circuits have to be designed so that the fault detection in such circuits becomes easier.

2.1 TEST GENERATION FOR COMBINATIONAL LOGIC CIRCUITS

Several distinct test generation methods have been developed over the years for combinational circuits. These methods are based on the assumptions that a circuit is nonredundant and only a single stuck-at fault is present at any time.

2.1.1   Truth Table and Fault Matrix

The most straightforward method for generating tests for a particular fault is to compare the responses of the fault-free and the faulty circuit to all possible input combinations. Any input combination for which the output responses do not match is a *test* for the given fault.

Let the inputs to a combinational circuit be $x_1, x_2, ..., x_n$ and let Z be the output of the circuit. Let Z_a be the output of the circuit in the presence of the fault α. The test generation method starts with the construction of the truth tables of Z and Z_a. Then for each row of the truth table, $Z \oplus Z_a$ is computed; if the result is 1, the input combination corresponding to the row is a test for the fault.

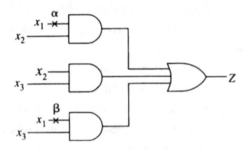

(a) A combinational circuit

x_1	x_2	x_3	z	z_α	z_β	$z + z_\alpha$	$z + z_\beta$
0	0	0	0	0	0	0	0
0	0	1	0	0	1	0	1
0	1	0	0	0	0	0	0
0	1	1	1	1	1	0	0
1	0	0	0	0	0	0	0
1	0	1	1	1	1	0	0
1	1	0	1	0	1	1	0
1	1	1	1	1	1	0	0

(b) Truth table for the fault-free and the faulty circuit

FIGURE 2.1: (a) A combinational circuit. (b) Truth table for the fault-free and faulty circuit.

As an example, let us consider the circuit shown in Figure 2.1a and assume that tests for faults α s-a-0 and β s-a-1 have to be derived. The truth table for the circuit is shown in Figure 2.1b, where column Z denotes the fault-free output, and Z_α and Z_β correspond to the circuit output in presence of faults α s-a-0 and β s-a-1, respectively. The tests for the faults are indicated as 1's in the columns corresponding to $Z \oplus Z_a$ and $Z \oplus Z_\beta$. Thus, the test for α s-a-0 is $x_1 x_2 x_3 = 110$, and the test for β s-a-1 is $x_1 x_2 x_3 = 001$. For all other input combinations, the output of the fault-free circuit is the same as the output in the presence of the fault; consequently, they are not tests for α s-a-0 and β s-a-1.

The minimum number of tests required to detect a set of faults in a combinational circuit can be obtained from a *fault matrix*. The columns in a fault matrix list the single faults to be tested, and the rows indicate the tests. A fault matrix for the circuit of Figure 2.2a is shown in Figure 2.2b. A 1 at the intersection of the ith row and the jth column indicates that the fault corresponding to the jth column can be detected by the ith test. As can be seen from Figure 2.2b, a fault matrix is identical to a *prime implicant chart* used in logic minimization. Thus, the problem of finding the minimum number of tests is the same as the problem of finding the minimum number of *prime implicants* (i.e., rows) so that every column has a 1 in at least one row. In Figure 2.2b, rows 110, 101, and 111 are equivalent (i.e., each test detects the same faults as the other two); hence, 101 and 111 can be omitted. Furthermore, row 000 covers row 100 and row 001 covers row 011; thus, rows 100 and 011 can be omitted. Elimination of rows 100, 101, 011, and 111 yields the minimal test set as shown in Figure 2.2c. These four tests detect all of the six faults under consideration.

It is obvious from this example that the fault matrix approach to test generation is not practicable when the number of input variables is large. We now discuss some alternative techniques developed to solve test generation problems.

2.1.2   Path Sensitization

The basic principle of the path sensitization method is to choose some path from the origin of the fault to the circuit output. As mentioned earlier, a path is sensitized if the inputs to the gates along the path are assigned values such that the effect of the fault can be propagated to the output [1].

To illustrate, let us consider the circuit shown in Figure 2.3 and assume that line α is s-a-1. To test for α, both G_3 and C must be set at 1. In addition, D and G_6 must be set at 1 so that $G_7 = 1$ if the fault is absent. To propagate the fault from G_7 to the circuit output f via G_8 requires the output of G_4 to be 1. This is because if $G_4 = 0$, the output f will be forced to be 1, independent of the value of gate G_7. The process of propagating the effect of the fault from its original location to the circuit output is known as the *forward trace*.

The next phase of the method is the *backward trace*, in which the necessary signal values at the gate outputs specified in the forward trace phase are established. For example, to set G_3 at 1, A must

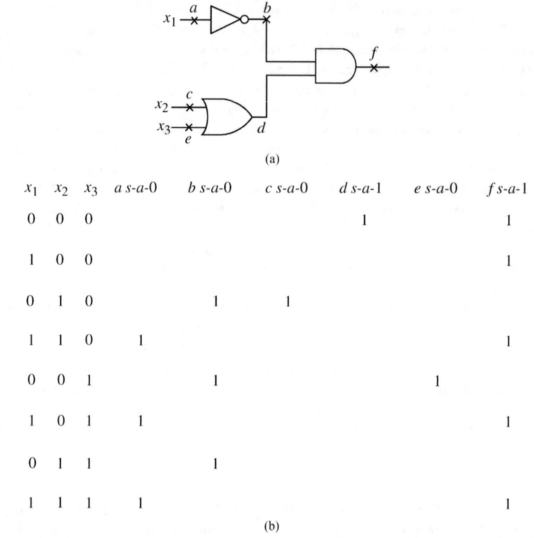

(a)

x_1	x_2	x_3	a s-a-0	b s-a-0	c s-a-0	d s-a-1	e s-a-0	f s-a-1
0	0	0				1		1
1	0	0						1
0	1	0		1	1			
1	1	0	1					1
0	0	1		1			1	
1	0	1	1					1
0	1	1		1				
1	1	1	1					1

(b)

x_1	x_2	x_3	a s-a-0	b s-a-0	c s-a-0	d s-a-1	e s-a-0	f s-a-1
0	0	0				1		1
0	1	0		1	1			
1	1	0	1					1
0	0	1		1			1	

(c)

FIGURE 2.2: (a) Circuit under test. (b) Fault matrix. (c) Minimal test set.

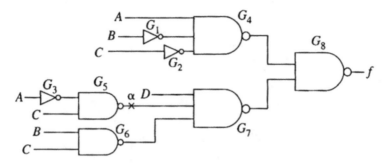

FIGURE 2.3: A combinational circuit with line α s-a-1.

be set at 0, which also sets G_4=1. In order for G_6 to be at 1, B must be set at 0; note that G_6 cannot be set at 1 by making C=0 because this is inconsistent with the assignment of C in the forward trace phase. Therefore, the test $ABCD$=0011 detects the fault α s-a-1, since the output f will be 0 for the fault-free circuit and 1 in the presence of the fault.

In general, a test pattern generated by the path sensitization method may not be unique. For example, the fault α s-a-0 in the circuit of Figure 2.4 can be detected by ABC=01- or 0-0. In the first test, C is unspecified, and B is unspecified in the second test. An unspecified value in a test indicates that the test is independent of the corresponding input.

The main drawback of the path sensitization method is that only one path is sensitized at a time. This does not guarantee that a test will be found for a fault even if one exists. As an example, let us derive a test for the fault α s-a-0 in Figure 2.5 [2]. To propagate the effect of the fault along the path G_2–G_6–G_8 requires that B, C, and D should be set at 0. In order to propagate the fault through G_8, it is necessary to make G_4=G_5=G_7=0. Since B and D have already been set to 0, G_3 is 1, which makes G_7=0. To set G_5=0, A must be set to 1; as a result, G_1=0, which with B=0 will make G_4=1. Therefore, it is not possible to propagate the fault through G_8. Similarly, it is not possible to sensitize the path G_2–G_5–G_8. However, A=0 sensitizes the two paths simultaneously and also

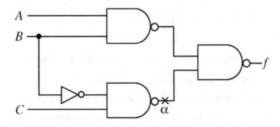

FIGURE 2.4: Circuit with α s-a-0.

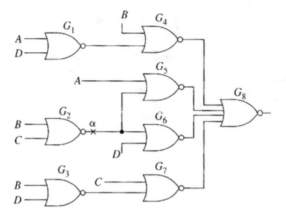

FIGURE 2.5: Three-level NOR logic circuit.

makes $G_4=0$. Thus, two inputs to G_8 change from 0 to 1 as a result of the fault α s-a-0, while the remaining two inputs remain fixed at 0. Consequently, $ABCD=0000$ causes the output of the circuit to change from 1 to 0 in the presence of α s-a-0 and is the test for the fault.

This example shows the necessity of sensitizing more than one path in deriving tests for certain faults and is the principal idea behind the D-algorithm.

2.1.3   D-Algorithm

The *D-algorithm* is guaranteed to find a test if one exists for detecting a fault [3]. It uses a cubical algebra for automatic generation of tests. Three types of cubes are considered:

1. Singular cube;
2. Propagation D-cube;
3. Primitive D-cube of a fault.

Singular cube. A singular cube corresponds to a *prime implicant* of a function. Figure 2.6 shows the singular cubes for the two-input NOR function; x's or blanks are used to denote that the position may be either 0 or 1.

Propagation D-cube. D-cubes represent the input/output behavior of the good and the faulty circuit. The symbol D may assume 0 or 1. $\overline{D}$ takes on the value opposite to D (i.e., if $D=1$, $\overline{D}=0$ and if $D=0$, $\overline{D}=1$). The definitions of D and $\overline{D}$ could be interchanged, but they should be consistent throughout the circuit. Thus, all D's in a circuit imply the same value (0 or 1) and all $\overline{D}$'s will have the opposite value.

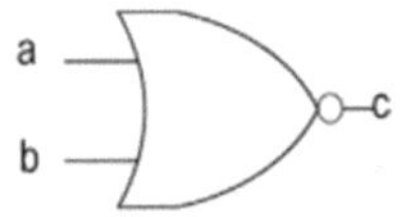

a	b	c
0	0	1
x	1	0
1	x	0

FIGURE 2.6: Singular cubes for the two-input NOR function.

The propagation D-cubes of a gate are those that cause the output of the gate to depend only on one or more of its specified inputs. Thus, a fault on a specified input is propagated to the output. The propagation D-cubes for a two-input NAND gate are:

$$
\begin{array}{ccc}
a & b & f \\
1 & D & \overline{D} \\
D & 1 & \overline{D} \\
D & D & \overline{D}
\end{array}
$$

The propagation D-cubes $1D\overline{D}$ and $D1\overline{D}$ indicate that if one of the inputs of the NAND gate is 1, the output is the complement of the other. $DD\overline{D}$ propagates multiple input changes through the NAND gate. Propagation D-cubes of a gate can be constructed by intersecting its singular cubes with output values. The intersection rules are as follows:

$$
\begin{aligned}
& 0 \cap 0 = 0 \cap x = x \cap 0 = 0 \\
& 1 \cap 1 = 1 \cap x = x \cap 1 = 1 \\
& x \cap x = x \\
& 1 \cap 0 = D \\
& 0 \cap 1 = \overline{D}
\end{aligned}
$$

For example, the propagation D-cube of a three-input NOR gate can be formed as shown in Figure 2.7.

	a	b	c	f
c_1	0	0	0	1
c_2	x	x	1	0
c_3	x	1	x	0
c_4	1	x	x	0

(a) Singular covers of
the NOR gate

	a	b	c	f
$c_2 \cap c_1$	0	0	D	$\overline{D}$
$c_3 \cap c_1$	0	D	0	$\overline{D}$
$c_4 \cap c_1$	D	0	0	$\overline{D}$

(b) Propagation D-cube of
the NOR gate

FIGURE 2.7: (a) Singular covers of the NOR gate. (b) Propagation D-cube of the NOR gate.

Primitive D-cube of a fault. The primitive D-cube of a fault (pdcf) is used to specify the existence of a given fault. It consists of an input pattern which shows the effect of a fault on the output of the gate. For example, if the output of the NOR gate shown in Figure 2.6 is s-a-0, the corresponding pdcf is:

$$
\begin{array}{ccc}
a & b & f \\
0 & 0 & D
\end{array}
$$

Here, D is interpreted as being 1 if the circuit is fault-free and is 0 if the fault is present. The pdcf's for the NOR gate output s-a-1 are:

$$
\begin{array}{ccc}
a & b & f \\
1 & x & \overline{D} \\
x & 1 & \overline{D}
\end{array}
$$

The pdcf's corresponding to an output s-a-0 fault in a gate can be obtained by intersecting each singular cube having output 1 in the fault-free gate with each singular cube having output 0 in the faulty gate. Similarly, the pdcf's corresponding to an output s-a-1 fault can be obtained by intersecting each singular cube with output 0 in the fault-free gate, with each singular cube having output 1 in the faulty gate. The intersection rules are similar to those used for propagation D-cubes.

As an example, let us consider a three-input NAND gate with input lines a, b, and c and output line f. The singular cubes for the fault-free NAND gate are:

	a	b	c	f
c_1	0	x	x	1
c_2	x	0	x	1
c_3	x	x	0	1
c_4	1	1	1	0

Assuming the input line b is s-a-1, the singular cubes for the faulty NAND gate are:

	a	b	c	f
c_1'	0	x	x	1
c_2'	x	x	0	1
c_3'	1	x	1	0

Hence,

$$c_1 \cap c_3' = \overline{D}x1D \quad c_4 \cap c_1' = D11\overline{D}$$
$$c_2 \cap c_3' = 101D \quad c_4 \cap c_2' = 11D\overline{D}$$
$$c_3 \cap c_3' = 1x\overline{D}D$$

Therefore, the primitive D-cube of the b s-a-1 fault is $101D$. The pdcf's for all single stuck-at faults for the three-input NAND gate are:

a	b	c	f	Fault
0	x	x	D	f s-a-0
x	0	x	D	f s-a-0
x	x	0	D	f s-a-0
1	1	1	$\overline{D}$	f s-a-1
0	1	1	D	a s-a-1
1	0	1	D	b s-a-1
1	1	0	D	c s-a-1

Let us next consider how the various cubes described are used in the D-algorithm method to generate a test for a given fault. The test generation process consists of three steps:

Step 1. Select a pdcf for the given fault.

Step 2. Drive the D (or $\overline{D}$) from the output of the gate under test to an output of the circuit by successively intersecting the current *test cube* with the propagation D-cubes of successive gates. A test cube represents the signal values at various lines in the circuit during each step of the test generation process. The intersection of a test cube with the propagation D-cube of a successor gate results in a test cube.

Step 3. Justify the internal line values by driving back toward the inputs of the circuit, assigning input values to the gates so that a consistent set of circuit input values may be obtained.

Let us demonstrate the application of the D-algorithm by deriving a test for detecting the α s-a-1 fault in Figure 2.8a. The test generation process is explained in Figure 2.8b. As can be seen in Figure

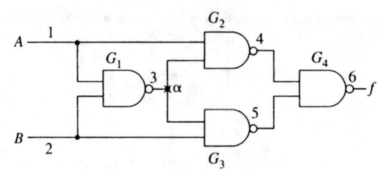

(a) Circuit example for D-Algorithm application

	1	2	3	4	5	6
Step 1: Select pdcf for α s-a-1.	1	1	$\overline{D}$	x	x	x
Step 2: Intersect the test cube with an appropriate propagation D-cube of G_2 (e.g., $1\overline{D}D$). (N.B., the parity of the D-cube is inverted.)	1	1	$\overline{D}$	D	x	x
Step 3: Intersect the test cube with the propagation D-cube $D1\overline{D}$ of G_4.	1	1	$\overline{D}$	D	1	$\overline{D}$
Step 4: Check that line 5 is at 1 from G_3 singular cubes.	1	#	$\overline{D}$	D	1	D

(b) Consistency operation unsuccessful.

	1	2	3	4	5	6
Step 1: Select pdcf for α s-a-1.	1	1	$\overline{D}$	x	x	x
Step 2: Intersect the test cube with the propagation cube $1\overline{D}D$ of G_2.	1	1	$\overline{D}$	D	x	x
Step 3: Intersect the test cube with the propagation cube $\overline{D}1D$ of G_3.	1	1	$\overline{D}$	D	D	x
Step 4: Intersect the test cube with the propagation cube $DD\overline{D}$ of G_1.	1	1	$\overline{D}$	D	D	$\overline{D}$

(c) Consistency operation not needed.

FIGURE 2.8: Derivation of test for α s-a-1.

2.8b, the consistency operation at step 4 terminates unsuccessfully because the output of G_3 has to be set to 1. This can be done only by making input $B=0$; however, B has already been assigned 1 in step 1. A similar problem will arise if D is propagated to the output via G_3 instead of G_2. The only way the consistency problem can be resolved is if the $\overline{D}$ output of G_1 is propagated to the output of the circuit via both G_2 and G_3 as shown in Figure 2.8c. No consistency operation is needed in this case, and the test for the given fault is $AB=11$. This test also detects the output of G_2 s-a-0, the output of G_3 s-a-0, and the output of G_4 s-a-1.

As a further example of the application of the D-algorithm, let us derive a test for the s-a-0 fault at the output of gate G_2 in the circuit shown in Figure 2.9a. The test derivation is as shown in Figure 2.9b. The test is $ABC=011$.

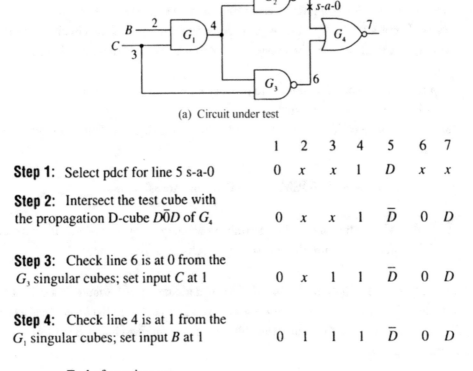

(a) Circuit under test

	1	2	3	4	5	6	7
Step 1: Select pdcf for line 5 s-a-0	0	x	x	1	D	x	x
Step 2: Intersect the test cube with the propagation D-cube $D\overline{0}D$ of G_4	0	x	x	1	$\overline{D}$	0	D
Step 3: Check line 6 is at 0 from the G_3 singular cubes; set input C at 1	0	x	1	1	$\overline{D}$	0	D
Step 4: Check line 4 is at 1 from the G_1 singular cubes; set input B at 1	0	1	1	1	$\overline{D}$	0	D

End of consistency

(b) Consistency operation successful

FIGURE 2.9: Sample application of D-algorithm.

2.1.4 PODEM

PODEM is an enumeration algorithm in which all input patterns are examined as tests for a given fault [4]. The search for a test continues until the search space is exhausted or a test pattern is found. If no test pattern is found, the fault is considered to be undetectable. In D-algorithm, *line justification*, i.e., line values assigned during the backtracking toward the inputs of the circuit, allows assignments on any internal lines. In PODEM, backtracking is allowed on primary inputs only, thus reducing the number of backtracks. PODEM consists of six steps:

> *Step 1.* Assume all primary inputs are x, which are unassigned. Determine an initial *objective*; an objective is defined by a logic (0 or 1) value referred to as *objective logic level*. The initial objective is to select a logic value so that the fault to be detected is sensitized.
>
> *Step 2.* Select a primary input and assign a logic value that has good likelihood of satisfying the initial objective.
>
> *Step 3.* Propagate forward the value at the selected primary input in conjunction with X's at the rest of the primary inputs by using the five-valued logic 0, 1, X, D, and $\overline{D}$.
>
> *Step 4.* If it is a test, a D or a $\overline{D}$ is propagated to the output of the circuit, exit; otherwise, assign the complement of the previous value to the primary input and determine whether it is a test.
>
> *Step 5.* Assign a 0 or a 1 to one more primary input, and go to step 4 to check whether the resulting combination is a test.
>
> *Step 6.* Continue with steps 4 and 5 until a test is found, or the fault is found to be undetectable.

The main differences between PODEM and D-algorithm are as follows:

1. In PODEM, backtracking is allowed only on primary inputs not on any internal line.
2. PODEM does not require the consistency check operation.

Let us illustrate the application of PODEM by deriving a test for fault l s-a-1 in the circuit shown in Figure 2.10. Since a test for fault l s-a-1 is to be derived, the initial objective is to set l to 0. Either B or C can be assigned 1 to satisfy the objective. Assuming we choose B to be at 1, the result of the forward propagation is:

A	B	C	l	m	n	p	F
X	1	X	$\overline{D}$	0	X	X	X

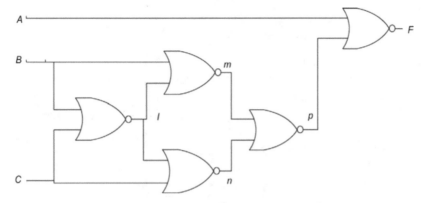

FIGURE 2.10: Circuit under test.

The next objective is to propagate $\overline{D}$ (or D) through n to output F. This can be done by assigning proper logic value to input C. Suppose we set C to 1, this results in the following:

$$
\begin{array}{cccccccc}
A & B & C & l & m & n & p & F \\
X & 1 & 1 & \overline{D} & 0 & 0 & X & X
\end{array}
$$

This will block the propagation of D because n is forced to 0. However if C is assigned 1, D is propagated through n:

$$
\begin{array}{cccccccc}
A & B & C & l & m & n & p & F \\
X & 1 & 0 & \overline{D} & 0 & D & \overline{D} & X
\end{array}
$$

The final objective is to propagate $\overline{D}$ (or D) to output F. This can be done by assigning proper logic value i.e. 0 to input A.

$$
\begin{array}{cccccccc}
A & B & C & l & m & n & p & F \\
0 & 1 & 0 & \overline{D} & 0 & D & \overline{D} & D
\end{array}
$$

Thus, $ABC=010$ is the test for l s-a-1.

As a further example, let us consider the derivation of a test for fault α s-a-0 in the circuit shown in Figure 2.11. The initial objective is to set the output of gate A to logic 1; that is, the objective logic level is 1 on net. Both primary inputs x_1 and x_2 drive gate A. Let us first assign 0 to x_1 as shown below:

$$
\begin{array}{cccccccccccc}
1 & 2 & 3 & 4 & 5 & 6 & 7 & 8 & 9 & 10 & 11 & 12 \\
0 & X & X & X & X & X & X & X & X & X & X & X
\end{array}
$$

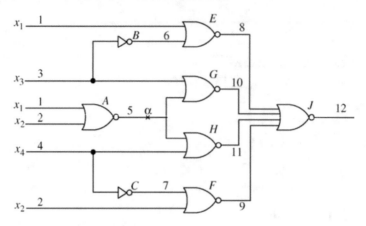

FIGURE 2.11: Test derivation using PODEM.

Since $x_1x_2x_3x_4=0XXX$ is not a test for the fault, we assign 0 to x_2 (the other primary input to A) which sets up D as the output of gate A:

$$
\begin{array}{cccccccccccc}
1 & 2 & 3 & 4 & 5 & 6 & 7 & 8 & 9 & 10 & 11 & 12 \\
0 & 0 & X & X & D & X & X & X & X & X & X & X
\end{array}
$$

Because the output of gate A, namely, net 5, is not X, it is necessary to find a gate closer to the primary output with D as its input and X as output. Both gates G and H satisfy the requirements. The selection of gate G and the subsequent assignment of 0 to primary input x_3 results in the following:

$$
\begin{array}{cccccccccccc}
1 & 2 & 3 & 4 & 5 & 6 & 7 & 8 & 9 & 10 & 11 & 12 \\
0 & 0 & 0 & X & D & 1 & X & 0 & X & \overline{D} & X & X
\end{array}
$$

It is clear that $x_1x_2x_3x_4=000X$ is not a test for the fault because the primary output is X. Gate J has $\overline{D}$ on input net 10 and X's on input nets 9 and 11. The initial objective is to set the objective net 12 to logic 1. The selection of net 9 as the next objective results in the assignment of 0 to primary input x_4:

$$
\begin{array}{cccccccccccc}
1 & 2 & 3 & 4 & 5 & 6 & 7 & 8 & 9 & 10 & 11 & 12 \\
0 & 0 & 0 & 0 & D & 1 & 1 & 0 & 0 & \overline{D} & \overline{D} & D
\end{array}
$$

Thus, the test for the fault α s-a-0 is $x_1x_2x_3x_4=0000$. The same test could be found for the fault by applying the D-algorithm; however, the D-algorithm requires substantial trial and error before the test is found. This is because of the variety of propagation paths and the attendant consistency operations that are required. For example, α s-a-0 has to be simultaneously propagated to the output via the paths *AGJ* and *AHJ*; propagation along either path individually will lead to inconsistency. This feature of the *D*-algorithm can lead to a waste of effort if a given fault is untestable. Thus, PODEM is more efficient than the *D*-algorithm in terms of computer time required to generate tests for combinational circuits.

2.1.5 FAN

The FAN algorithm is in principle similar to PODEM but is made more efficient by reducing the number of backtracks [5]. Several terms have to be defined before discussing the test generation process used by FAN. A *bound line* is a gate output that is part of a reconvergent fan-out loop. A line that is not bound is considered to be *free*. A *headline* is a free line that drives a gate that is part of a reconvergent fan-out loop. In Figure 2.12, for example, nodes *H*, *I*, and *J* are bound lines, *A* through *H* are free lines, and *G*, *H*, and *F* are headlines. Because by definition headlines are free lines, they can be considered as primary input lines and can always be assigned values arbitrarily. Thus, during a backtrack operation if a headline is reached, the backtrack stops; it is not necessary to reach a primary input to complete the backtrack.

FAN uses a technique called *multiple backtracks* to reduce the number of backtracks that must be made during the search process. For example, in Figure 2.13, if the objective is to set *H* at logic 1, PODEM would backtrack along one of the paths to the primary inputs. Suppose the backtrack is done via the path *H-E-C*, which will set *E* to 1. Because *E* is at 1, *C* will set to 0. However a 0 at *C* sets *F* to 1, *G* to 0, and *H* to 0. Because this assignment fails to achieve the desired objective, the backtrack process is performed via another path, for example, *H–G–F–C*, and the desired goal can be achieved. Thus, in PODEM, several backtracks may be necessary before the requirement of setting up a particular logic value on a line is satisfied. FAN avoids this waste of computation time by backtracking along multiple paths to the fan-out point. For example, if multiple backtrack is

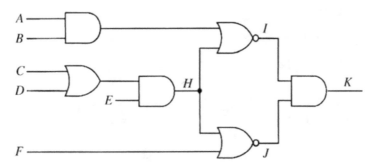

FIGURE 2.12: Illustration of bound line, free line and head line.

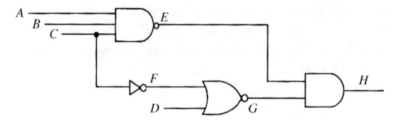

FIGURE 2.13: Multiple backtracks along H–E–C and H–G–F–C.

done via both H–E–C and H–G–F–C, the value at C can be set so that the value at H is justified. In PODEM, a logic value assigned to a primary input in order to achieve one objective may in turn result in the failure of satisfying another objective, thereby forcing a backtrack.

We illustrate the application of the FAN algorithm by deriving a test for the fault Z s-a-0 in Figure 2.14. First, the value D is assigned to the line Z and the value 1 to each of the inputs M and N. The initial objectives are to set M and N to 1. By the multiple backtrack, G and I are assigned 1 (note that instead of G and I, L could be assigned logic 1). Again, by the multiple backtrack, we have the final objectives $A=1$, $B=1$ and $E=1$, $F=1$. The assignment $A=1$, $B=1$ makes $J=1$, $M=1$, and the assignment $E=1$, $F=1$ makes $I=1$, $N=1$. Thus, the assignments $A=B=E=F=1$ constitute a test for the fault Z s-a-0. It is easy to see that if the first multiple backtracks stopped at L and the second multiple backtrack at H, the test for the fault would be $C=D=1$.

2.1.6 Delay Fault Detection

A delay fault in a combinational logic circuit can be detected only by applying a sequence of two test patterns. The first pattern, known as an *initialization pattern*, sets up the initial condition in a circuit

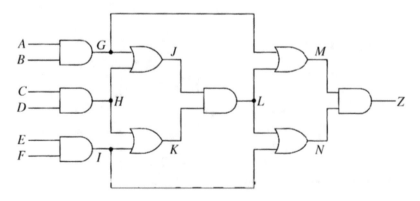

FIGURE 2.14: Circuit under test.

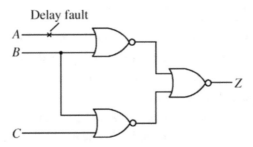

FIGURE 2.15: Detection of slow-to-rise delay fault.

so that the fault (slow-to-rise or slow-to-fall signal) at the input or output of a gate affect an output of the circuit. The second pattern, known as a *transition or propagation pattern*, propagates the effect of the activated transition to a primary output of the circuit.

To illustrate, let us consider a delay (slow-to-rise) fault at the input A of the circuit shown in Figure 2.15. The test for slow-to-rise fault consists of the initialization pattern ABC=001 followed by the transition pattern ABC=101. Similarly, the two pattern tests for a slow-to-fall delay fault at input A will be ABC=101, 001. Note that a slow-to-rise fault and a slow-to-fall fault correspond to a transient stuck-at-0 and transient stuck-at-1 fault, respectively.

To identify the presence of a delay fault in a combinational circuit, the hardware model shown in Figure 2.16 is frequently used in literature. The initialization pattern is first loaded into the input latches. After the circuit has stabilized, the transition pattern is clocked into the input latches by using $C1$. The output pattern of the circuit is next loaded into the output latches by setting the clock $C2$ at logic 1 for a period equal to or greater than the time required for the output pattern to be loaded into the latch and stabilize. The possible presence of a delay fault is confirmed if the output value is different from the expected value.

Delay tests can be classified into two groups: *nonrobust* and *robust* [6]. A delay fault is nonrobust if it can detect a fault in the path under consideration provided there are no delay faults along

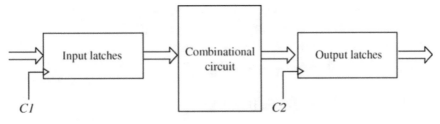

FIGURE 2.16: Hardware model for delay fault testing.

other paths. For example, the input vector pair (111, 101) can detect the slow-to-rise fault at *e* in Figure 2.17a as long as the path *b–d–f* does not have a delay fault. However, if there is a slow-to-fall fault at *d*, the output of the circuit will be correct for the input pair, thereby invalidating the test for the delay fault at *e*. Therefore, the test (111, 101) is nonrobust.

A delay test is considered to be robust if it detects the fault in a path independent of delay faults that may exist in other paths of the circuit. For example, let us assume a slow-to-fall delay fault at *d* in the path *a–c–d–f* of the circuit shown in Figure 2.17b. The input vector pair (01, 11) constitutes a robust test for the delay fault because the output of any gate on the other paths does not change when the second vector of the input pair is applied to the circuit. Thus, any possible delay fault in these paths will not affect the circuit output. Robust tests do not exist for many paths in large circuits [7, 8].

2.2   TESTING OF SEQUENTIAL CIRCUITS

Test generation for sequential circuits is extremely difficult because the behavior of a sequential circuit depends both on the present and on the past input values. The mathematical model of a

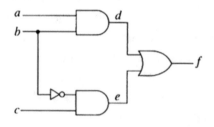

(a) Nonrobust test

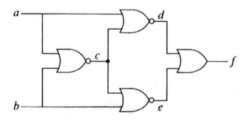

(b) Robust test

FIGURE 2.17: (a) Nonrobust test. (b) Robust test.

synchronous sequential circuit is usually referred to as a sequential machine or a finite state machine. Henceforth, a synchronous sequential circuit will be referred to as a sequential circuit.

Figure 2.18 shows the general model of a synchronous sequential circuit. As can be seen from the diagram, sequential circuits are basically combinational circuits with memory to remember past inputs. The combinational part of the circuit receives two sets of input signals: *primary* (coming from the external environment) and *secondary* (coming from the memory elements). The particular combination of secondary input variables at a given time is called the *present state* of the circuit; the secondary input variables are also known as *state variables*. If there are m secondary input variables in a sequential circuit, then the circuit can be in any one of 2^m different present states. The outputs of the combinational part of the circuit are divided into two sets. The primary outputs are available to control operations in the circuit environment, whereas the secondary outputs are used to specify the *next state* to be assumed by the memory. It takes an entire sequence of inputs to detect many of the possible faults in a sequential circuit.

Sequential circuits can be tested by checking that such a circuit functions as specified by its state table [9, 10]. This is an exhaustive approach and is practical only for small sequential circuits. The approach may be summarized as follows: Given the state table of a sequential circuit, find an input/output sequence pair (X, Z) such that the response of the circuit to X will be Z if and only if the circuit is operating correctly. The application of this input sequence X and the observation of the response, to see if it is Z, is called a *checking experiment*; the sequence pair (X, Z) is referred to as a *checking sequence*.

The derivation of checking sequence for a sequential circuit is based on the following assumptions:

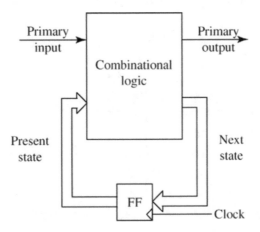

FIGURE 2.18: Synchronous sequential circuit.

1. The circuit is fully specified and *deterministic*. In a deterministic circuit the next state is determined uniquely by the present state and the present input.
2. The circuit is *strongly connected*; that is, for every pair of states q_i and q_j of the circuit, there exists an input sequence that takes the circuit from q_i to q_j.
3. The circuit in the presence of faults has no more states than those listed in its specification. In other words, the presence of a fault will not increase the number of states.

To design checking experiments, it is necessary to know the initial state of the circuit which is determined by a *homing sequence* or a *distinguishing sequence*. An input sequence is said to be a homing sequence for a sequential circuit if the circuit's response to the sequence is always sufficient to determine uniquely its final state. For an example, consider the state table of a circuit shown in Figure 2.19. It has a homing sequence 101, for, as indicated in Figure 2.20, each of the output sequences that might result from the application of 101 is associated with just one final state. A homing sequence need not always leave a machine in the same final state; it is only necessary that the final state can be identified from the output sequence.

A distinguishing sequence is an input sequence that, when applied to a sequential circuit, will produce a different output sequence for each choice of initial state. For example, 101 is also a distinguishing sequence for the circuit shown in Figure 2.19. As shown in Figure 2.20, the output sequence that the machine produces in response to 101 uniquely specifies its initial state. Every distinguishing sequence is also a homing sequence because the knowledge of the initial state and the input sequence is always sufficient to determine uniquely the final state as well. On the other hand, not every homing sequence is a distinguishing sequence. For example, the circuit specified by the state table of Figure 2.21a has a homing sequence 010. As shown in Figure 2.21b, the output sequence produced in response to 010 uniquely specifies the final state of the circuit but cannot

Present state	Input	
	$x = 0$	$x = 1$
A	C,1	D,0
B	D,0	B,1
C	B,0	C,1
D	C,0	A,0
	Next state, output	

FIGURE 2.19: State table for a circuit.

Initial state	Output sequence	Final state
A	0 0 1	C
B	1 0 0	A
C	1 0 1	B
D	0 1 1	C

FIGURE 2.20: Response of the sequential circuit to homing sequence 101.

distinguish between the initial states C and D. Every reduced sequential circuit possesses a homing sequence, whereas only a limited number of sequential circuits have distinguishing sequences.

At the start of an experiment, a circuit can be in any of its n states. In such a case, the *initial uncertainty* regarding the state of the circuit is the set that contains all the states of the circuit. A collection of states of the circuit that is known to contain the present state is referred to as the *uncertainty*. The uncertainty of a circuit is thus any subset of the state of the machine. For example, the state table of Figure 2.19 can initially be in any of its four states; hence, the initial uncertainty is $(ABCD)$. If an input 1 is applied to the circuit, the successor uncertainty will be (AD) or (BC) depending on whether the output is 0 or 1, respectively. The uncertainties $(C)(DBC)$ are the 0-successors of $(ABCD)$. A successor tree, which is defined for a specified circuit and a given initial uncertainty, is a structure that displays graphically the x_i-successor uncertainties for every possible input sequence x_i.

A collection of uncertainties is referred to as an *uncertainty vector*, the individual uncertainties contained in the vector are called the components of the vector. An uncertainty vector, the components of which contain a single state each, is said to be a *trivial uncertainty vector*. An uncertainty

Present state	Input $x = 0$	Input $x = 1$	Present state	Response to 010			Final state
A	B,0	D,0	A	0	0	0	A
B	A,0	B,0	B	0	0	1	D
C	D,1	A,0	C	1	0	1	D
D	D,1	C,0	D	1	0	1	D
(a)			(b)				

FIGURE 2.21: (a) State table with homing sequence 101. (b) Response to the homing sequence.

vector, the components of which contain either single states or identical repeated states, is said to be a homogeneous uncertainty vector. For example, the vectors $(AA)(B)(C)$ and $(A)(B)(A)(C)$ are homogeneous and trivial, respectively.

A homing sequence is obtained from the homing tree; a homing tree is a successor tree in which a node becomes terminal if one of the following conditions occurs:

1. The node is associated with an uncertainty vector, the nonhomogeneous components of which are associated with the same node at a preceding level.
2. The node is associated with a trivial or a homogeneous vector.

The path from the initial uncertainty to a node in which the vector is trivial or homogeneous defines a homing sequence.

A distinguishing tree is a successor tree in which a node becomes terminal if one of the following conditions occurs:

1. The node is associated with an uncertainty vector, the nonhomogeneous components of which are associated with the same node at a preceding level.
2. The node is associated with an uncertainty vector containing a homogeneous nontrivial component.
3. The node is associated with a trivial uncertainty vector.

The path from the initial uncertainty to a node associated with a trivial uncertainty defines a distinguishing sequence. As an example, the homing sequence 010 is obtained as shown in Figure 2.22 by applying the terminal rules to state table of Figure 2.21a. The derivation of the distinguishing sequence 101 for state table of Figure 2.19 is shown in Figure 2.23.

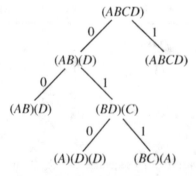

FIGURE 2.22: Homing tree for the state table of Figure 2.21.

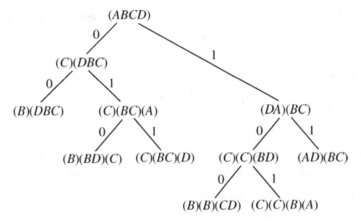

FIGURE 2.23: Distinguishing tree for the state table of Figure 2.19.

During the design of checking experiments, it is often necessary to take the circuit into a predetermined state, after the homing sequence has been applied. This is done with the help of a *transfer sequence*, which is the shortest input sequence that takes a machine from state S_i to state S_j. The procedure is an adaptive one, because the transfer sequence is determined by the response of the homing sequence. As an example, let us derive a transfer sequence that will take state table of Figure 2.19 from state B to state C. To accomplish this, we assume that the circuit is in state B. We form the transfer tree as shown in Figure 2.24; it can be seen from the successor tree that the shortest transfer sequence that will take the machine from state B to state C is 00.

2.2.1   Designing Checking Experiments

Basically, the purpose of a checking experiment is to verify that the state table of a sequential circuit accurately describes its behavior. If during the execution of the experiment the circuit produces a

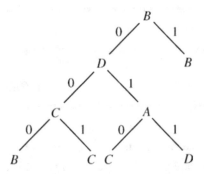

FIGURE 2.24: Transfer tree.

response that is different from the correctly operating circuit, the circuit is definitely faulty. Such experiments can be used only to determine whether or not something is wrong with a circuit; it is not possible to conclude from these experiments what is wrong with the circuit.

A checking experiment can be designed for a strongly connected sequential circuit provided it has at least one distinguishing sequence. The checking experiment can be divided into three phases:

1. *Initialization phase.* During the initialization phase, the circuit under test is taken from an unknown initial state to a fixed state. A reduced, strongly connected circuit can be maneuvered into some fixed states by the following method:
   a. Apply a homing sequence to the circuit and identify the current state of the circuit.
   b. If the current state is not s, apply a transfer sequence to move the circuit from the current state to s.
2. *State identification phase.* During this phase, an input sequence is applied so as to cause the circuit to visit each of its states and display its response to the distinguishing sequence.
3. *Transition verification phase.* During this phase, the circuit is made to go through every state transition; each state transition is checked by using the distinguishing sequence.

Although these three phases are distinct, in practice, the subsequences for state identification and transition verification are combined whenever possible in order to shorten the *length* of the experiment. The length is the total number of input symbols applied to the circuit during the execution of an experiment; it is a measure of efficiency of the experiment.

2.3   TEST GENERATION USING THE CIRCUIT STRUCTURE AND THE STATE TABLE

A test generation technique for sequential circuits based on the concept of path sensitization used in combinational circuit test generation has been proposed in Ref. [11]. This technique takes into account both the structure and the state table of a sequential circuit. It is assumed that the circuit under test has a reset state. A test sequence is applied to the circuit with the *reset state* as the starting state. The test generation process consists of the following steps:

1. Generate a test vector for the assumed single stuck-at fault such that the effect of the fault is propagated to the primary outputs or to the secondary outputs, i.e., the outputs of the flip-flops. Each primary output as well as each secondary output is considered an independent output of a combinational circuit. A test vector for a fault is identified as an *excitation vector*, and the present state part of an excitation vector is called the *excitation state*.

2. Derive an input sequence to take the circuit from the reset state to the excitation state; this input sequence is called the *justification sequence*. Obviously, a justification sequence is not necessary if the excitation state part of a test vector is the reset state. If the effect of a fault can be propagated to the primary outputs by the derived test vector, and the justification sequence can take the faulty circuit from the reset to the excitation state, then the test vector is valid. However, if the test vector can propagate the effect of the fault only to the outputs of the flip-flops, i.e., the next state is different from the one expected, the following step is also necessary for successful test generation for the assumed fault.

3. Derive an input sequence such that the last bit in the sequence produces a different output for the fault-free and the faulty states of the circuit under test. Such an input sequence is called the *differentiating sequence*.

A test sequence for the fault under test is obtained by concatenating the justification sequence, the excitation vector, and the differentiating sequence. This test sequence is simulated in the presence of the fault to check if the fault is detected. If the fault is not detected, the differentiating sequence is not valid. Moreover, a valid differentiating sequence cannot be obtained if the fault-free and the faulty states are equivalent in the fault-free sequential circuit.

Let us illustrate the technique by deriving a test sequence for the fault α s-a-0 in the sequential circuit shown in Figure 2.25a; the state table of the fault-free circuit is shown in Figure 2.25b. The state are encoded as shown in Figure 2.25c.

A partial test vector for the fault is first derived:

$$x \quad y_1 \quad y_2$$
$$0 \quad 1 \quad -$$

The value of y_2, has to be chosen such that the effect of the fault is propagated to the primary output. If $y_2=0$, the fault is neither propagated to the output nor does it affect the next state. On the other hand, $y_2=1$ affects the next state variable y_1. Therefore, the excitation vector for the fault is:

$$x \quad y_1 \quad y_2$$
$$0 \quad 1 \quad 1$$

and the excitation state is 11 (i.e., state *C*).

Next, we derive the justification sequence that can take the fault-free machine from the reset state *A* to the excitation state *C*. It can be verified from the state table that the justification sequence needs to contain a single bit only, which is 0. Thus, the test sequence derived so far consists of:

$$0 \quad 0$$

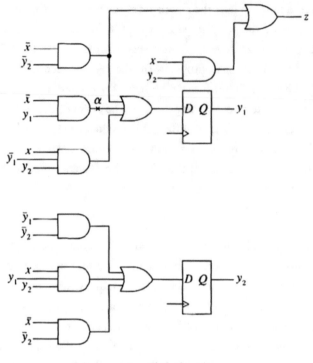

(a) A sequential circuit

	x = 0	x = 1
A	C,1	B,0
B	A,0	D,1
C	D,0	B,1
D	C,1	A,0

(b) State Table

	y_1	y_2
A	0	0
B	0	1
C	1	1
D	1	0

(c) State Encoding

FIGURE 2.25: (a) A sequential circuit. (b) State table. (c) State encoding.

and the corresponding fault-free output sequence and the next states are:

$$\begin{matrix} 0 & & 0 \\ A \to & C \to & D \\ 1 & & 0 \end{matrix}$$

In the presence of the assumed stuck-at fault, the next state/output sequence is:

$$\begin{array}{ccc} 0 & 0 & \\ A \rightarrow C \rightarrow A \\ 1 & 0 & \end{array}$$

Because the output sequence is the same as in the fault-free case, the fault is not detectable. However, the final state in the presence of the fault is A instead of expected D; in other words, the effect of the fault α propagated only to the outputs of the flip-flops. Therefore, a differentiating sequence that produces a different output sequence for state A and state D has to be concatenated with the previously derived test sequence. The differentiating sequence is derived as follows:

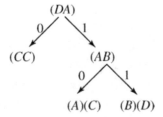

Either 10 or 11 can be used as the differentiating sequence. The choice of 11 results in the following test sequence for the assumed stuck-at fault in the circuit of Figure 2.25a:

$$0 \quad 0 \quad 1 \quad 1$$

The corresponding next state/output sequence for the fault-free circuit is:

$$\begin{array}{ccccc} 0 & 0 & 1 & 1 & \\ A \rightarrow C \rightarrow D \rightarrow A \rightarrow B \\ 1 & 0 & 0 & 0 & \end{array}$$

and for the faulty circuit:

$$\begin{array}{ccccc} 0 & 0 & 1 & 1 & \\ A \rightarrow C \rightarrow A \rightarrow B \rightarrow D \\ 1 & 0 & 0 & 1 & \end{array}$$

Thus, the fault is detected by the derived test sequence.

REFERENCES

[1] Armstrong, D. B., "On finding a nearly minimal set of fault detection tests for combinational logic nets," IEEE *Trans. Electron. Comput., 66–73 (* February 1966).

[2] Schneider, R. R., "On the necessity to examine D-chains in diagnostic test generation," *IBM Jour. Res. Dev.*, 114 (January 1967).

[3] Roth, J. P., "Diagnosis of automata failures: A calculus and a method," *IEEE Trans. Comput.*, 278–91 (July 1966).

[4] Goel, P., "An implicit enumeration algorithm to generate test for combinational logic circuits," *IEEE Trans. Comput.*, 215–22 (May 1981).

[5] Fujiwara, H. and T. Shimono, "On the acceleration of test generation algorithms," *IEEE Trans. Comput.*, 1137–44 (December 1983).

[6] Park, E. and M. Mercer, "Robust and nonrobust tests for path delay faults in a combinational circuit," *Proc. Intl. Test Conf.*, 1027–34 (1987).

[7] Reddy, S. M., C. Li, and S. Patil, "An automatic *test* pattern generator for the detection of path delay faults," *Proc. IEEE Intl. Conf. CAD*, 284–7 (November 1987).

[8] Schulz, M. H., K. Fuchs, and F. Fink, "Advanced automatic test pattern generation techniques for path delay faults," *Proc. 19th IEEE Intl. Fault-Tolerant Comput.* Symp., 44–51 (June 1989). doi:10.1109/FTCS.1989.105541

[9] Kohavi, Z., *Switching and Finite Automata Theory*, Chap. 13, McGraw-Hill (1970).

[10] Hennie, F. C., *Finite State Models for Logical Machines*, Chap 3, John Wiley (1968).

[11] Ghose, A., S. Devadas, and A. R. Newton, "Test generation and verification for highly sequential circuits," *IEEE Trans. CAD*, 652–67 (May 1961).

·   ·   ·   ·

CHAPTER 3

Design for Testability

The phrase *design for testability* refers to how a circuit is either designed or modified so that the testing of the circuit is simplified. Several techniques have been developed over the years for improving the testability of logic circuits. These can be categorized into two categories: *ad hoc* and *structured*.

The ad hoc approaches simplify the testing problem for a specific design and cannot be generalized to all designs. On the other hand, the structured techniques are generally applicable to all designs.

3.1   AD HOC TECHNIQUES

One of the simplest ways of improving the testability of a circuit is to provide more tests and control points. Test points are, in general, used to observe the response at a node inside the circuit, whereas control points are utilized to control the value of an internal node to any desired value, 0 or 1. For example, in the circuit shown in Figure 3.1a, the fault α s-a-0 is undetectable at the circuit output.

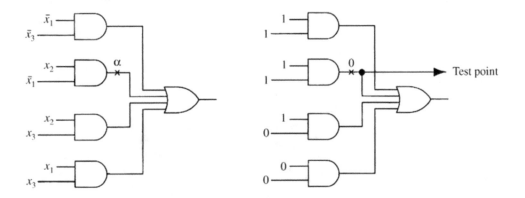

(a) Circuit with undetectable fault α s-a-0 (b) Fault detected when $x_1x_2x_3$ =010 is applied

FIGURE 3.1: (a) Circuit with undetectable fault α s-a-0. (b) Fault detected when $x_1x_2x_3$=010 is applied.

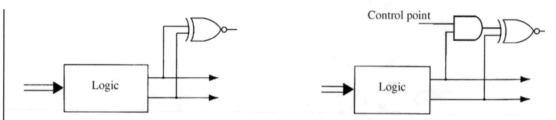

FIGURE 3.2: (a) EX-NOR gate not testable. (b) EX-NOR gate easily testable.

By incorporating a test point at node α as shown in Figure 3.1b, the input combination 010 or 011 can be applied to detect the fault.

The usefulness of adding a control point can be appreciated from the circuit shown in Figure 3.2a. If the output of the EX-NOR gate in the circuit is always 1, indicating that both the outputs of the logic block are the same, it is not possible to say whether the EX-NOR gate is operating correctly or not. If a control point is added to the circuit, as shown in Figure 3.2b, the input of the EX-NOR gate and hence the operation of the circuit can be controlled via the added point. During the normal operation of the circuit, the control point is set at logic 1. To test for an s-a-1 fault at the output of the EX-NOR gate, the control point is set at logic 0 and an input combination that produces logic 1 at the outputs has to be applied.

Another way of improving the testability of a particular circuit is to insert multiplexers in order to increase the number of internal nodes that can be controlled or observed from the external points. For example, the fault α s-a-0 in Figure 3.1a can also be detected by incorporating a 2-to-1 multiplexer as shown in Figure 3.3. When the test input (i.e., the select input of the multiplexer) is

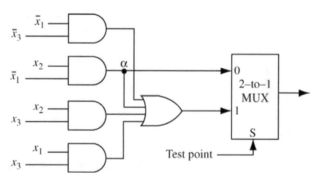

FIGURE 3.3: Use of multiplexer to enhance testability.

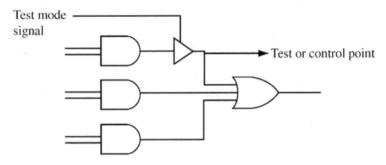

FIGURE 3.4: Use of tristate drivers to improve testability.

at logic 1, the output of the circuit is transferred to the output of the multiplexer. On the other hand, if the control input is at logic 0 and the input combination 010 or 011 is applied to the circuit, the state of node α can be observed at the multiplexer output.

A different way of accessing internal nodes is to use tristate drivers as shown in Figure 3.4. A test mode signal could be used to put the driver into the high-impedance state. In this mode, the input of the OR gate can be set to logic 0 or logic 1 from an external point. When the driver is enabled, the same external point becomes a test point.

The test mode signals required by the added components, such as multiplexers, tristate drivers, etc., cannot always be applied via external points, because it is often not practicable to have many such points. To reduce the number of external points, a *test state register* may be included in the

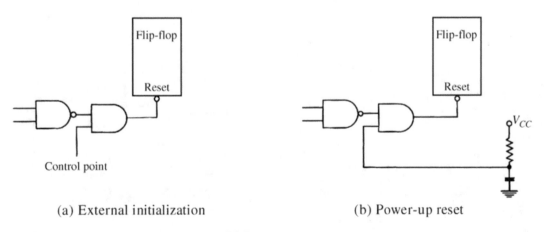

(a) External initialization   (b) Power-up reset

FIGURE 3.5: (a) External initialization. (b) Power-up reset.

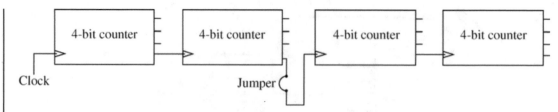

FIGURE 3.6: Breaking up of a counter chain.

circuit. This could in fact be a shift register that is loaded and controlled by just a few pins. The testability hardware in the circuit can then be controlled by the parallel outputs of the shift register.

Frequently, flip-flops, counters, shift registers, and other memory elements assume unpredictable states when power is applied, and they must be set to known states before testing can begin. Ideally, all memory elements should be reset from external points (Figure 3.5a). Alternatively, a power-up reset may be added to provide internal initialization (Figure 3.5b).

A long counter chain presents another test problem. For example, the counter chain in Figure 3.6 requires thousands of clock pulses to go through all the states. One way to avoid this problem is to break up the long chains into smaller chains by filling jumpers to them; the jumpers can be removed during testing. A tristate driver can function as a jumper in this case. The input of the tristate driver is connected to the clock, and the output to the clock input of the second counter chain. When the control input of the tristate driver is disabled, the clock is disconnected from the second counter chain; thus, this chain can be tested separately from the first chain.

A feedback loop is also difficult to test because it hides the source of the fault. The source can be located by breaking the loop and bringing both lines to external points that are shown during

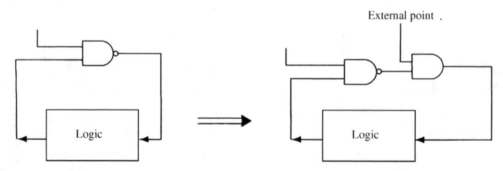

FIGURE 3.7: Breaking a feedback loop by using an extra gate.

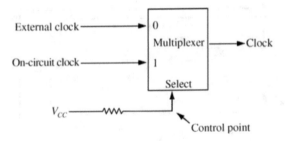

FIGURE 3.8: Replacement of on-circuit clock.

normal operation. When not shorted, the separate lines provide a control point and a test point. An alternative way of breaking a feedback loop is to add to the feedback path a gate that can be interrupted by a signal from a control point (Figure 3.7).

On-circuit clock oscillators should be disconnected during test and replaced with an external clock. The external clock can be single-stepped to check the logic values at various nodes in the circuit during the fault diagnosis phase. Figure 3.8 shows how the onboard clock can be replaced by an external one.

3.2   SCAN-PATH TECHNIQUE FOR TESTABLE SEQUENTIAL CIRCUIT DESIGN

The testing of sequential circuits is complicated because of the difficulties in setting and checking the states of the memory elements. These problems can be overcome by modifying the design of a general sequential circuit so that it has the following properties [1].

1.   The circuit can easily be set to any desired internal state.
2.   It is easy to find a sequence of input patterns such that the resulting output sequence will indicate the internal state of the circuit.

The basic idea is to add an extra input c to the memory excitation logic in order to control the mode of a circuit. When $c=0$, the circuit operates in its normal mode, but when $c=1$, the circuit enters into a mode in which the elements are connected together to form a shift register. This facility is incorporated by inserting a double-throw switch, i.e., a 2-to-1 multiplexer in each input lead of every memory element. All these switches are grouped together, and the circuit can operate in either its *normal* or *shift register* mode. Figure 3.9 shows a sequential circuit using D flip-flops; the circuit is modified as shown in Figure 3.10. Each of the double-throw switches may be realized as

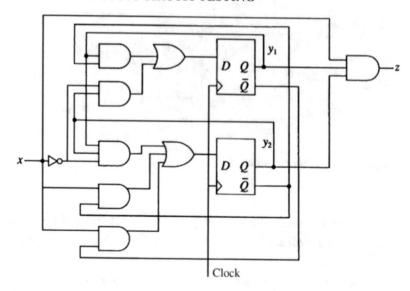

FIGURE 3.9: A sequential circuit.

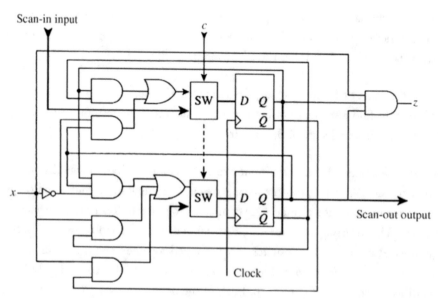

FIGURE 3.10: Modified sequential circuit.

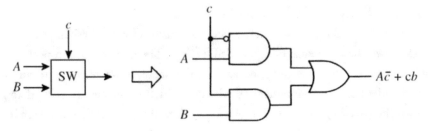

FIGURE 3.11: A realization of the double-throw switch.

indicated in Figure 3.11. One additional input connection to the modified circuit is required to sup-
ply the signal c to control all the switches.

In the shift register mode, the first flip-flop can be set directly from the primary inputs
(scan-in inputs) and the output of the last flip-flop can be directly monitored on the primary output
(scan-out output). This means that the circuit can be set to any desired state via the scan-in inputs
and that the internal state can be determined via the scan-out output.

The procedure for testing the circuit is as follows:

1. Set $c=1$ to switch the circuit to shift register mode.
2. Check operation as a shift register by using scan-in inputs, scan-out output, and the
    clock.
3. Set the initial state of the shift register.
4. Set $c=0$ to return to normal mode.
5. Apply test input pattern to the combinational logic.
6. Set $c=1$ to return to shift register mode.
7. Shift out the final state while setting the starting state for the next test.
8. Go to step 3.

With this procedure, a considerable proportion of the actual testing time is spent in setting
the state, an operation that requires a number of clock pulses equal to the length of the shift register.
This time may be decreased by forming several short shift registers rather than a single long one; the
time needed to set or read the state would then be equal to the length of the longest shift register.
The extent to which the number of shift registers can be increased is determined by the number of
input and output connections available to be used to drive and sense the shift registers.

The main advantage of the scan-path approach is that a sequential circuit can be transformed
into a combinational circuit, thus making test generation for the circuit relatively easy. Besides, very
few extra gates or pins are required for this transformation.

3.3   LEVEL-SENSITIVE SCAN DESIGN

One of the best known and the most widely practiced methods for synthesizing testable sequential circuits is IBM's level-sensitive scan design (LSSD) [2–5]. The *level-sensitive* aspect of the method means that a sequential circuit is designed so that the steady-state response to any input state change is independent of the component and wire delays within the circuit. Also, if an input state change involves the changing of more than one-input signal, the response must be independent of the order in which they change. These conditions are ensured by the enforcement of certain design rules, particularly pertaining to the clocks that evoke state changes in the circuit. *Scan* refers to the ability to shift into or out of any state of the circuit.

3.3.1   Clocked Hazard-Free Latches

In LSSD, all internal storage is implemented in hazard-free polarity-hold latches. The polarity-hold latch has two-input signals as shown in Figure 3.18a. The latch cannot change state if $C=0$. If C is set to 1, the internal state of the latch takes the value of the excitation input D. A flow table for this sequential circuit, along with an excitation table and a logic implementation, is shown in Figure 3.18b, 3.18c, and 3.18d, respectively.

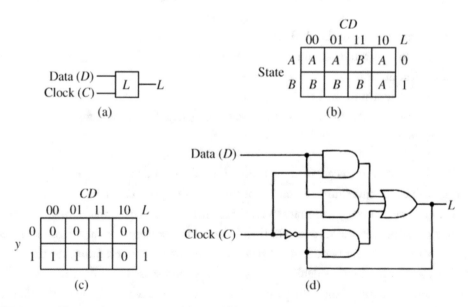

FIGURE 3.12: Hazard-free polarity-hold latch: (a) symbolic representation; (b) flow table; (c) excitation table; (d) logic implementation (Reprinted from Ref. [2], © 1978).

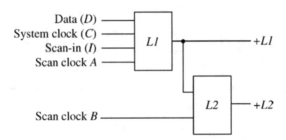

FIGURE 3.13: Polarity-hold SRL (Reprinted from Ref. [2], © 1978).

The clock signal *C* will normally occur (change from 0 to 1) after the data signal *D* has be-
came stable at either 1 or 0. The output of the latch is set to the new value of the data signal at the
time the clock signal occurs. The correct changing of the latch does not depend on the rise or fall
time of the clock signal, but only on the clock signal being 1 for a period equal to a greater than the
time required for the data signal to propagate through the latch and stabilize.

A shift register latch (SRL) can be formed by adding a clocked input to the polarity-hold
latch *L1* and including a second latch *L2* to act as intermediate storage during shifting (Figure
3.13). As long as the clock signals *A* and *B* are both 0, the *L1* latch operates exactly like a polarity-
hold latch. Terminal *I* is the scan-in input for the SRL and +*L2* is the output. The logic implementation of the SRL is shown in Figure 3.14. When the latch is operating as a shift register data from
the preceding stage are gated into the polarity-hold switch via *I*, through a change of the clock *A*
from 0 to 1. After *A* has changed back to 0, clock *B* gates the data in the latch *L1* into the output
latch *L2*. Clearly, *A* and *B* can never both be 1 at the same time if the SRL is to operate properly.

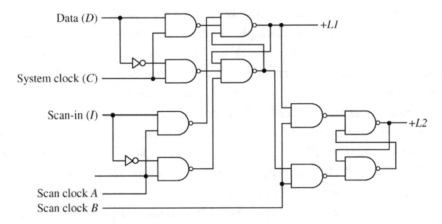

FIGURE 3.14: Logic for shift-register latch (Reprinted from Ref. [2], © 1978).

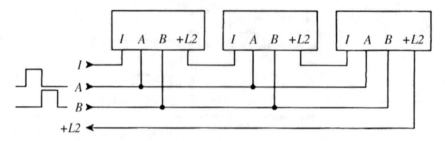

FIGURE 3.15: Linkage of three SRLs (Reprinted from Ref. [2], © 1978).

The SRLs can be interconnected to form a shift register as shown in Figure 3.15. The input I and the output $+L2$ are stung together in a loop, and the clocks A and B are connected in parallel.

A specific set of design rules has been defined to provide level-sensitive logic subsystems with a scannable design that would aid testing:

Rule 1. Use only hazard-free polarity-hold latches as memory elements.

Rule 2. The latches must be controlled by nonoverlapping clocks.

Rule 3. Clock signals must be applied via primary inputs.

Rule 4. Clocks may not feed the data inputs to memory elements either directly or through combinational logic.

Rule 5. Test sequences must be applied via a primary input.

3.3.2   Double-Latch and Single-Latch LSSD

A sequential logic circuit that is level-sensitive and also has the scan capability is called a Level Sensitive Scan Design (LSSD). Figure 3.16 depicts a general structure for an LSSD system, known as a *double-latch design* in which all system outputs are taken from the $L2$ latch. In this configuration, each SRL operates in a master–slave mode. Data transfer occurs under system clock and scan clock B during normal operation and under scan clock A and scan clock B during scan-path operation. Both latches are therefore required during system operation.

In the single-latch configuration, the combinational logic is partitioned into two disjoint sets, Combl and Comb2 (Figure 3.17). The system clocks used for SRLs in Combl and Comb2 are denoted by Clock 1 and Clock 2, respectively; they are nonoverlapping. The outputs of the SRLs in Combl are fed back as secondary variable inputs to Comb2, and vice versa. This configuration uses the output of latch $L1$ as the system output; the $L2$ latch is used only for shifting. In other words, the $L2$ latches are redundant and represent the overhead for testability.

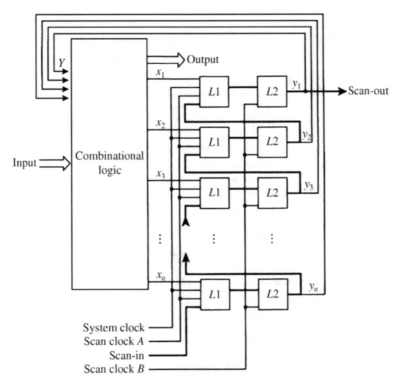

FIGURE 3.16: Double-latch LSSD (Adapted from Ref. [6]).

3.4   RANDOM ACCESS SCAN TECHNIQUE

The design methods discussed in Sections 3.2 and 3.3 use sequential access scan-in/scan-out techniques to improve testability; that is, all flip-flops are connected in series during testing to form a shift register or registers. In an alternative approach, known as *random access scan*, each flip-flop in a logic circuit is selected individually by an address for control and observation of its state [7]. The basic memory element in a random access scan-in/scan-out network is an *addressable latch*. The circuit diagram of an addressable latch is shown in Figure 3.18. A latch is selected by X–Y address signals, the state of which can then be controlled and observed through scan-in/scan-out lines. When a latch is selected and its scan clock goes from 0 to 1, the scan data input is transferred through the circuit to the scan data output, where the inverted value of the scan data can be observed. The input on the DATA line is transferred to the latch output Q during the negative transition (1 to 0) of the clock. The scan data out lines from all latches are then AND-gated to produce the chip scan-out signal: the scan-out line of a latch remains at logic 1 unless the latch is selected by the X–Y signals.

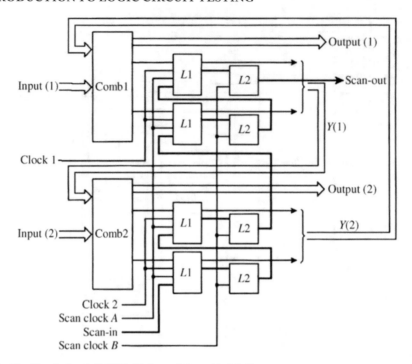

FIGURE 3.17: Single-latch LSSD (Adapted from Ref. [6]).

A different type of addressable latch—the set–reset type—is shown in Figure 3.19. The "clear" signal clears the latch during its negative transition. Prior to scan-in operation, all latches are cleared. Then, a latch is addressed by the X–Y lines and the preset signal is applied to set the latch state.

The basic mode1 of a sequential circuit with random access scan-in/scan-out feature is shown in Figure 3.20. The X- and Y-address decoders are used to access an addressable latch like a cell in

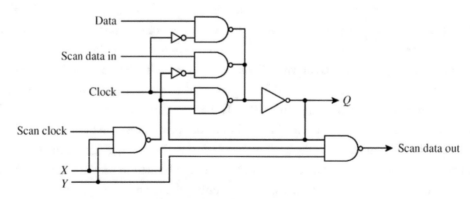

FIGURE 3.18: An addressable latch (Reprinted from Ref. [7], © 1980).

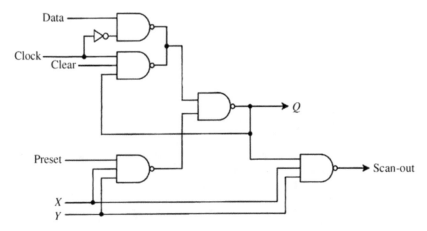

FIGURE 3.19: Set/reset type addressable latch (Reprinted from Ref. [7], © 1980).

random access memory. A tree of AND gates is used to combine all scan-out signals. Clear input of all latches are tied together to form a master reset signal. Preset inputs of all latches receive the same scan-in signal gated by the scan clock however, only the latch accessed by the X–Y address is affected.

The test procedure of a sequential circuit with random access scan-in/scan-out feature is as follows:

1. Set test input to all test points.
2. Apply the master reset signal to initialize all memory elements.
3. Set scan-in address and data and then apply the scan clock.
4. Repeat step 3 until all internal test inputs are scanned in.
5. Clock once for normal operation.
6. Check states of the output points.
7. Read the scan-out states of all memory elements by applying appropriate.
8. X–Y signals.

The random access scan-in/scan-out technique has several advantages:

1. The observability and controllability of all system latches are allowed.
2. Any point in a combinational circuit can be observed with one additional gate and one address per observation point.
3. A memory array in a logic circuit can be tested through a scan-in/scan-out circuit. The scan address inputs are applied directly to the memory array. The data input and the write-

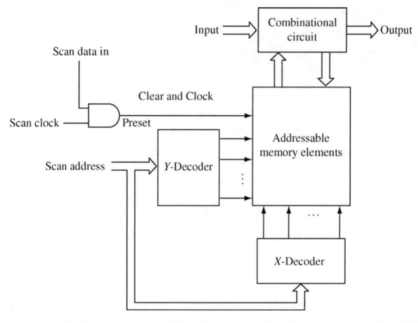

FIGURE 3.20: Sequential circuit design with addressable latches (Reprinted from Ref. [7], © 1980).

enable input of the array receive the scan data and the scan clock, respectively. The output of the memory array is AND-gated into the scan-out tree to be observed.

The technique has also a few disadvantages:

1. Extra logic in the form of two address gates for each memory element, plus the address decoders and output AND trees, result in 3–3 gates overhead per memory element.
2. Scan control, data, and address pins add up to 10–20 extra pins. By using a serially loadable address counter, the number of pins can be reduced to around 6.
3. Some constraints are imposed on the logic design such as the exclusion of asynchronous latch operation.

3.5   PARTIAL SCAN

In full scan, all flip-flops in a circuit are connected into one or more shift registers; thus, the states of a circuit can be controlled and observed via the primary input and outputs, respectively. In partial scan, only a subset of the circuit flip-flops is included in the scan chain in order to reduce the overhead associated with full scan design [6]. Figure 3.21 shows a structure of partial scan design. This has two separate clocks: a system clock and a scan clock. The scan clock controls only the scan flip-flops. Note that the scan clock is derived by gating the system clock with the scan-enable signal; no external clock is necessary. During the normal mode of operation, i.e., when the scan-enable signal is at logic 0, both scan and nonscan flip-flops update their states when the system clock is applied. In the scan mode operation, only the state of the shift register (constructed from the scan flip-flops) is shifted one bit with the application of the scan flip-flop; the nonscan flip-flops do not change their states.

The disadvantage of two-clock partial scan is that the routing of two separate locks with small skews is very difficult to achieve. Also, the use of a separate scan clock does not allow the testing of the circuit at its normal operating speed.

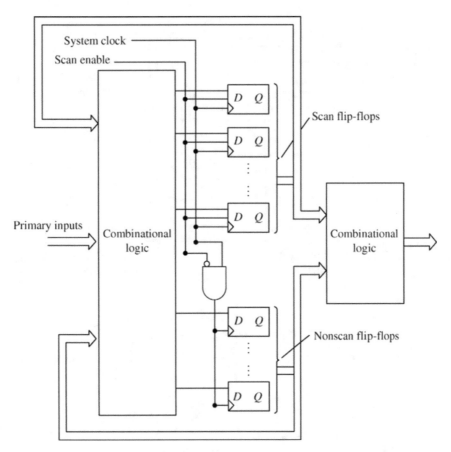

FIGURE 3.21: Partial scan using two clocks.

A partial scan scheme that uses the system clock as the scan clock is shown in Figure 3.22. [8]. Both scan and nonscan flip-flops move to their next states when the system clock is applied. A test sequence is derived by shifting data into the scan flip-flops. This data together with contents of nonscan flip-flops constitute the starting state of the test sequence. The other patterns in the sequence are obtained by single-bit shifting of the contents of scan flip-flops, which form part of the required circuit states. The remaining bits of the states, i.e., the contents of the scan flip-flops are determined by the functional logic. Note this form of partial scan scheme allows only a limited number of valid next states to be reached from the starting state of the test sequence. This may limit the fault coverage obtained by using the technique.

The selection flip-flops to be included in the partial scan is done by heuristic methods. It has been shown that the fault coverage in a circuit can be significantly increased by including 13–23% of the flip-flops flops in the partial scan [9].

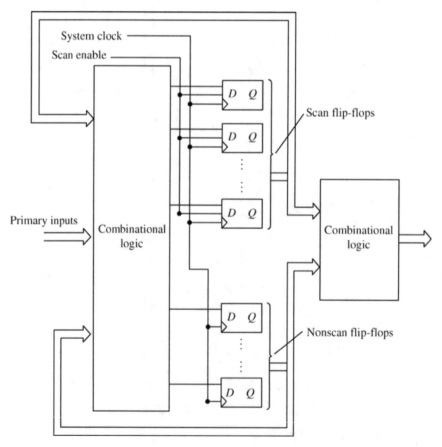

FIGURE 3.22: Partial scan using the system clock.

3.6 TESTABLE SEQUENTIAL CIRCUIT DESIGN USING NONSCAN TECHNIQUES

Full and partial scan techniques improve the controllability and observability of flip-flops in sequential circuit, and therefore the test generation for such circuits is considerably simplified. However, a scan-based circuit cannot be tested at its normal speed, because test data have to be shifted in and out via the scan path.

Figure 3.23 shows the configuration of a testable sequential circuit that employs nonscan flip-flops [10]. The original sequential circuit is augmented with a *test mode* input. If this input is at logic

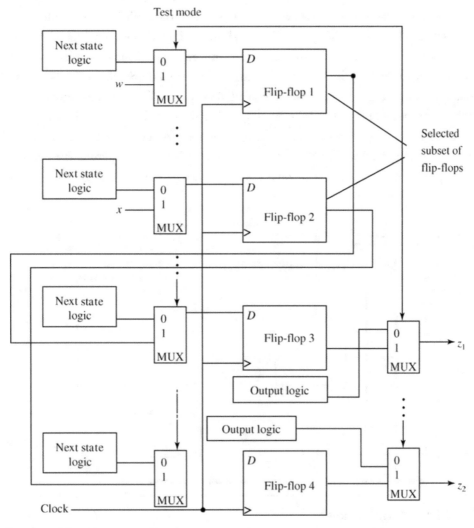

FIGURE 3.23: Testable sequential circuit.

1, each flip-flop from a selected subset of flip-flops in the circuit is connected directly to a primary input. This enables data to be shifted directly into these flip-flops from the primary inputs; at the same time, previous data from these flip-flops are shifted into the flip-flops that do not belong to the subset. The contents of the flip-flops in the selected subset are available on the primary outputs when the test mode signal is at logic 1. To illustrate, let us consider the two-input and two-output sequential circuit with four flip-flops shown in Figure 3.23. The primary inputs w and x are used to shift data into selected flip-flops 1 and 2, respectively, when the *test mode* signal is at logic 1. The previous contents of flip-flops 1 and 2 are shifted into flip-flops 3 and 3, respectively, while the contents of flip-flops 3 and 3 are observable via primary outputs z_1 and z_2.

Several techniques similar to that in Ref. [10] have been proposed in Ref. [11] to enhance controllability and observability in sequential circuits so that the testing of the circuits can be done at normal speed. The controllability of a circuit is improved by selecting a subset of flip-flops such that the number of cycles in the circuit's state diagram is minimized and each selected flip-flop can be loaded directly from one primary input line during the test mode. Such flip-flops are identified as *controllable* flip-flops. Figure 3.24 shows a sequential circuit modified for enhanced controllability. When the test input is at logic 0, all the flip-flops in a circuit are driven by their original next state logic. In other words, the circuit operates in its normal mode. When the test input is at logic 1, each controllable flip-flop is driven by a primary input line via a 2-to-1 multiplexer. The observability is improved by selecting a set of internal nodes that are untestable because faults at such nodes cannot

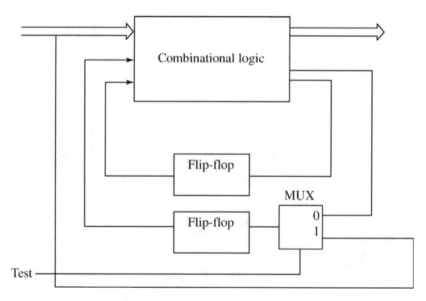

FIGURE 3.24: Enhanced controllability.

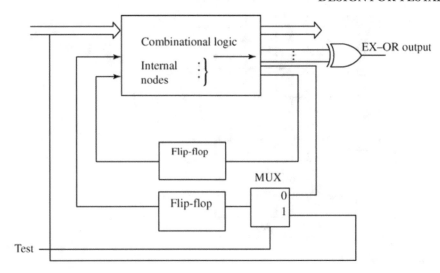

FIGURE 3.25: Observability enhancement.

be propagated to the primary outputs. The signals at these nodes are compressed by an EX-OR tree, the output of which is available on an additional output line as shown in Figure 3.25.

3.7   CROSSCHECK

The CrossCheck approach incorporates test circuitry into the basic cells used to implement VLSI circuits [12, 13]. This is done by connecting the output of a cell to the drain of a pass transistor. The gate of the transistor is connected to a probe line P and the source to a sense line S, as shown in Figure 3.26. The output of the cell can be observed via the S-line by controlling the probe line P. In other words, the controllability and the observability of the cells can be guaranteed.

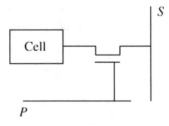

FIGURE 3.26: Basic cell.

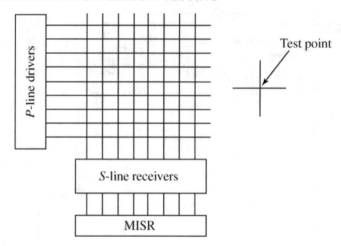

FIGURE 3.27: CrossCheck grid.

This approach can be used to enhance testability of VLSI chips by using cells with Cross-Check test points to implement logic. The individual test points are routed into an orthogonal grid of probe and sense lines as shown in Figure 3.27.

Appropriate test patterns are first applied to the circuit under test via the primary inputs. The probe lines are then enabled, allowing logic values at internal test points to be transferred to the sense lines. These values are stored in a parallel-in/serial-out shift register. Alternatively, a multiple-input signature register (MISR) can be used to compress the test values into a signature. It should be clear from this discussion that the CrossCheck approach allows parallel scanning of test points.

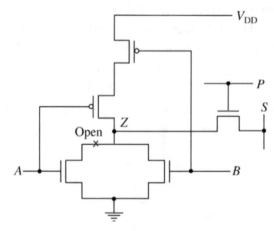

FIGURE 3.28: A two-input CMOS NOR gate with a stuck-open fault.

The CrossCheck testability approach simplifies the detection of stuck-open faults in logic cells. To illustrate, let us consider a two-input NOR gate augmented by a transistor having a *P*-line and an *S*-line (Figure 3.28). Let us assume the NOR gate has a stuck-open fault as shown in Figure 3.28. The fault can be activated by applying the test pattern *AB*=10. The sense line is precharged to a weak logic 1. When the probe line *P* is set at logic 1, the logic value on the *S*-line will be 0 in the absence of the assumed stuck-open fault; otherwise, the logic value will be 1. Thus, a stuck-open fault can be detected with a single test pattern.

The CrossCheck approach enhances the controllability of sequential circuits by using a D flip-flop implementation called *cross-controlled latch* (CCL). A CCL consists of a master–slave edge-triggered flip-flop augmented by two transistors t_1 and t_2 (Figure 3.29). Transistor t_1 is controlled by a test-write-enable ($\overline{\text{TWE}}$) signal. When $\overline{\text{TWE}}$ =1, and also probe line *P* and the clock are set at logic 1, the value on node *Y* is observable on the sense line *S* via transistor t_2; this value is also observable at the *Q* output. When the $\overline{\text{TWE}}$ signal is set at logic 0, transistor t_1 is turned off and the feedback path is disabled. A CCL functions as a D flip-flop when the $\overline{\text{TWE}}$ signal is at logic 1. The overhead due to the extra transistors is around 3%; also there is a slight increase in the minimum clock pulse width. On the plus side, the setting of flip-flops to specific values can be done in parallel; thus the test generation process for sequential circuits may be faster than in scan-based circuits.

The CrossCheck approach allows high observability of test points in a circuit. This, in conjunction with the enhanced controllability resulting from the use of CCLs, makes the CrossCheck approach a powerful design for testability technique. The drawback of the approach is that the

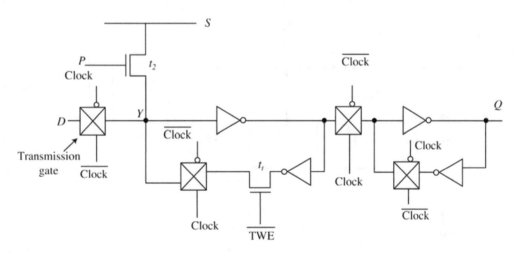

FIGURE 3.29: Cross-controlled latch.

scanning of test points introduces additional delay in the test application time. Thus, as in the scan technique, it is not possible to perform testing at the normal speed of the circuit.

3.8   BOUNDARY SCAN

Boundary scan methodology was initially developed to control and observe the input and output pins of chips used in assembling a system on a printed circuit board (PCB), it is now used to simplify testing of a system-on-a-chip (SOC). An international consortium, the Joint Test Action Group (JTAG), proposed a boundary scan architecture that was adopted by the IEEE Standards Board as IEEE Std. 1149.1 [14]. This architecture provides a single serial scan path through the input/output pins of individual chips on a board. The scan path is created by connecting the normal inputs/outputs of the chip logic to the input/output pins of the chip through *boundary scan cells*.

Figure 3.30 shows an implementation of a boundary scan cell. The operation of boundary scan cells is controlled by the test access pat (TAP), which has four inputs: test clock (TCK), test mode select (TMS), test data input (TDI), and test data output (TDO). During normal operation, data at the input (Data in) is transferred to the internal logic via the second multiplexer by setting Mode=0; the data can also stored in the first flip-flop by setting Shift=0 and clocking the flip-flop. During the test mode, data at the TDI is scanned in by setting Shift=1 and clocking the first flip-flop; the scanned data is available at TDO. The data captured in first flip-flop during the normal or the scan mode can be transferred to the output (Data out) by setting Mode =1 and clocking the *Update clock*.

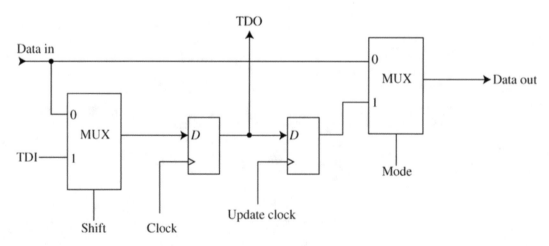

FIGURE 3.30: Boundary scan cell. (Taken from Ref. 14)

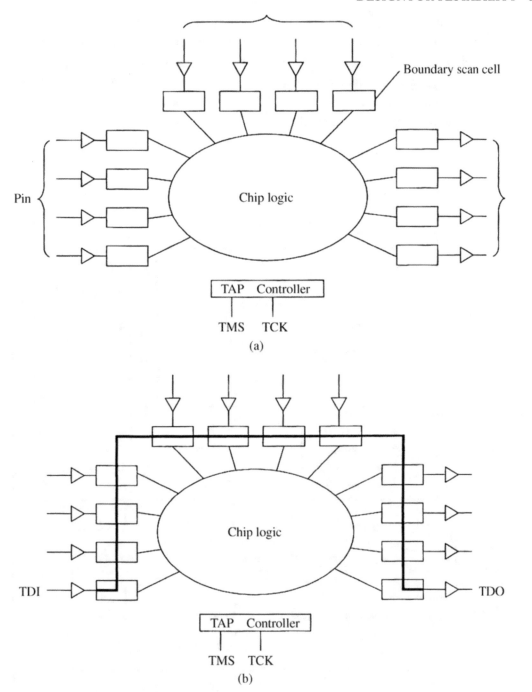

FIGURE 3.31: (a) Chip with boundary scan cells placed next to the pins. (b) Scan cells interconnected to form boundary scan register. [Taken from Ref. 14]

A chip with boundary scan cells placed next to the pins is shown in Figure 3.31a. These cells can be interconnected to form a continuous shift register (the boundary scan path) around the border of the chip. This shift register, as shown in Figure 3.31b, can be used to shift, apply, or capture test data. During normal operation of a chip, the boundary scan cells are transparent, and data applied at the input pins flows directly to the circuit. The corresponding output response of the circuit flows directly to the output pins of the chip.

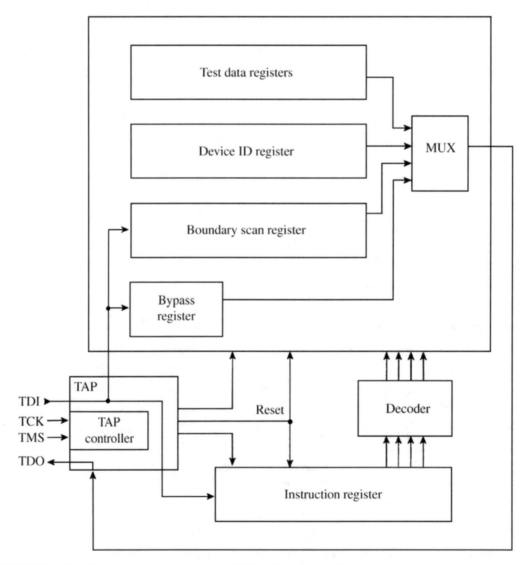

FIGURE 3.32: Boundary scan architecture. [Taken from Ref. 14]

During the scan mode test patterns are shifted in via the TDI pin, and the captured responses are shifted out via the TDO pin. Thus, the controllability and observability of the pins of a chip can be achieved without making the pins physically accessible. In addition to the TAP, the boundary scan architecture includes a controller for the TAP, an instruction register (IR), and a group of test data registers (TDRs).

Figure 3.32 shows a JTAG boundary scan architecture. The TAP controller is a finite state machine that generates various control signals—update, shift, or capture data—required in the boundary scan architecture. The state transitions in the TAP controller occur on the rising edge of the clock pulse at the TCK pin. The IR is a serial-in/serial-out shift register; the contents of the shift register are stored in a parallel output latch. Once the contents of the shift register, i.e., the current instruction, is loaded into the output latch, it can only be changed when the TAP controller is in either update-IR or test-logic-reset state. In the update-IR state, the newly shifted instruction is loaded into the output latch, whereas in the test-logic-reset state, all test logic is reset and the chip performs its normal operation. The TAP controller can be put in the test-logic-reset state from any other state by holding the TMS signal at logic 1 and by applying the TCK pulse at least five times.

The boundary scan architecture, in addition to the boundary scan register, contains another DR called the *bypass register*. The bypass register is only 1 bit wide. As can be seen in Figure 3.32,

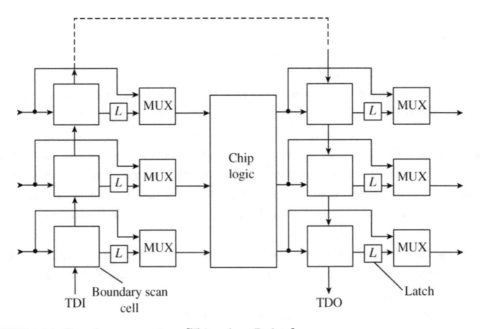

FIGURE 3.33: Boundary scan register. [Taken from Ref. 14]

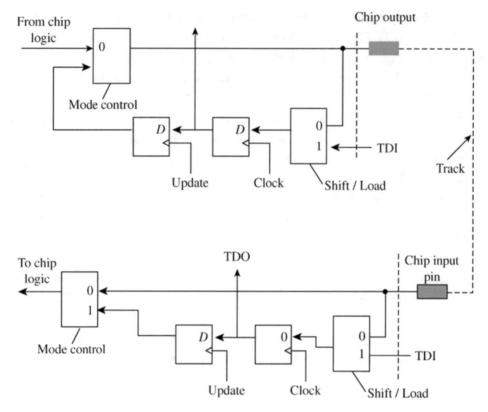

FIGURE 3.34: External test. [Taken from Ref. 14]

data at the TDI of a chip can be moved to its TDO through the bypass register in one clock cycle, rather than the multiple clock cycles required to shift data along the length of the scan path. Thus, the bypass register is extremely useful if only a small subset of all chips on a board need to be tested. Optionally, a device identification register may be included in a chip together with the TDRs. It is a 32-bit parallel-in/serial-out shift register and is loaded with data that identify the chip manufacture's name, part number, and version number when the TAP controller is in capture-data register (DR) state. The normal operation of a chip is not affected if either the ID register or the bypass register is in use.

As mentioned previously, boundary scan cells can be interconnected to form a shift register. This shift register is identified as the *boundary scan register*, the structure of this register is shown in Figure 3.33. The outputs of all shift register stages are stored in a latch in order to prevent the change of information while data is being shifted in and out.

A boundary scan register can be configured to perform three types of tests: *external test, internal test*, and *sample test*. The external test is used to test interconnections for stuck-at and bridging faults. It is invoked by entering logic 0 into every cell of the IR. Test patterns are then shifted into the boundary scan register stages at the output pins of a chip via the TDI pin. These patterns arriving at the input pins of other chips are loaded into their boundary scan registers, and they are observed via the TDO pins (Figure 3.33).

The internal test allows individual chips in a system to be tested. During this code, test vectors are shifted into the boundary scan register of the chip under test via the TDI pin. The corresponding test responses are loaded into the boundary scan register and shifted out for verification. Thus, the internal test allows individual chips in a system to be tested. The sample test allows monitoring of data flowing into and out of a chip during its normal operation. This can aid an external tester in understanding the chip's performance (Figure 3.34).

REFERENCES

[1] Williams M. J. Y. and J. M. Angell, "Enhancing testability of large-scale integrated circuits via test points and additional logic," *IEEE Trans. Comput.*, 36–9 (January 1973).

[2] Eichelbarger, E. B. and T. W. Williams, "A logic system structure for LSI testability," *Proc. ACM/IEEE Design Automation Conf.*, 462–8 (June 1978).

[3] Godoy, H. C., G. B. Franklin, and P. S. Bottorff, "Automatic checking of logic design structure for compliance with testability ground rules," *Proc. ACM/IEEE Design Automation Conf.*, 469–78 (1978).

[4] Dasgupta, S., P. Goel, R. G. Walther, and T. W. Williams, "A variation of LSSD and its implementation on design and test pattern generation in VLSI," *Proc. Intl. Test Conf.*, 63–6 (1982).

[5] Williams, T. W. and K. P. Parker, "Design for testability—a survey," *IEEE Trans. Comput.*, 2–13 (January 1982).

[6] Cheng, K. T. and V. D. Agarwal, "A partial scan method for sequential circuits with feedback," *IEEE Trans. Comput.*, 544–8 (April 1990). doi:10.1109/12.54847

[7] Ando, H., "Testing of VLSI with random access scan," *Proc. COMPCON*, 50–2 (Spring 1980).

[8] Cheng, K. T., "Single clock partial scan," *IEEE Design Test Comput.*, 23–31 (Summer 1995).

[9] Trischler, E., "Testability analysis and complete scan path," *Proc. Intl. Conf. CAD*, 38–9 (1983).

[10] Reddy, S. M. and R. Dandapani, "Scan design using standard flip-flops," *IEEE Design Test Comput.*, 52–4 (February 1987).

[11] Chickemane, V., E. M. Rudnick, P. Banerjee, and J. H. Patel, "Non-scan design-for-testability-techniques for sequential circuits," *Proc. ACM/IEEE Design Automation Conf.*, 236241 (1993).

[12] Gheewala, T., "CrossCheck: A cell based VLSI testability solution," *Proc. ACM/IEEE Design Automation Conf.*, 706–9 (1989). doi:10.1145/74382.74509

[13] Chandra, I. C., T. Ferry, T. Gheewala, and K. Pierce, "ATPG based *on* a novel grid-addressable latch element," *Proc. ACM/IEEE Design Automation Conf.*, 282–6 (1991). doi:10.1145/127601.127681

[14] John Fluke Inc. "The ABCs of Boundary-Scan Test," 1991.

·   ·   ·   ·

CHAPTER 4

Built-in Self-Test

The task of testing a VLSI chip to guarantee its functionality is extremely complex and often very time consuming. In addition to the problem of testing the chips themselves, the incorporation of the chips into systems has caused test generation's cost to grow exponentially. A widely accepted approach to deal with the testing problem at the chip level is to incorporate built-in self-test (BIST) capability inside a chip. This increases the controllability and the observability of the chip, thereby making the test generation and fault detection easier. In conventional testing, test patterns are generated externally by using computer-aided design (CAD) tools. The test patterns and the expected responses of the circuit under test to these test patterns are used by an automatic test equipment (ATE) to determine if the actual responses match the expected ones. On the other hand, in BIST, the test pattern generation and the output response evaluation are done on chip; thus, the use of expensive ATE machines to test chips can be avoided.

A basic BIST configuration is shown in Figure 4.1. The main function of the test pattern generator is to apply test patterns to the unit under test (assumed to be a multi-output combinational circuit). The resulting output patterns are transferred to the output response analyzer. Ideally, a BIST scheme should be easy to implement and must provide a high fault coverage.

4.1 TEST PATTERN GENERATION FOR BIST

Test pattern generation approaches for BIST schemes can be divided into four categories:

1. Exhaustive testing;
2. Pseudoexhaustive testing;
3. Pseudorandom testing;
4. Deterministic testing.

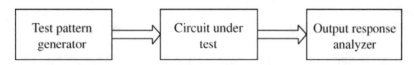

FIGURE 4.1: BIST configuration.

4.1.1 Exhaustive Testing

In the exhaustive testing approach, all possible input patterns are applied to the circuit under test. Thus, for an n-input combinational circuit, all possible 2^n patterns need to be applied. The advantage of this approach is that all nonredundant faults can be detected; however, any fault that converts the combinational circuit into sequential circuit, for example, a bridging fault, cannot be detected. The drawback of this approach is that when n is large, the test application time becomes prohibitive, even with high clock speeds. Thus, exhaustive testing is practical only for circuits with a limited number of inputs.

A modified form of exhaustive testing is *pseudoexhaustive testing* [1]. It retains the advantages of exhaustive testing while significantly reducing the number of test patterns to be applied. The basic idea is to partition the circuit under test into several sub-circuits such that each sub-circuit has few enough inputs for exhaustive testing to be feasible for it. This concept has been used in *autonomous design verification* technique proposed in Ref. [2]. To illustrate the use of the technique, let us consider the circuit shown in Figure 4.2a. The circuit is partitioned into two sub-circuits C_1 and C_2 as shown in Figure 4.2b. The functions of the two added control inputs MC1 and MC2 are as follows:

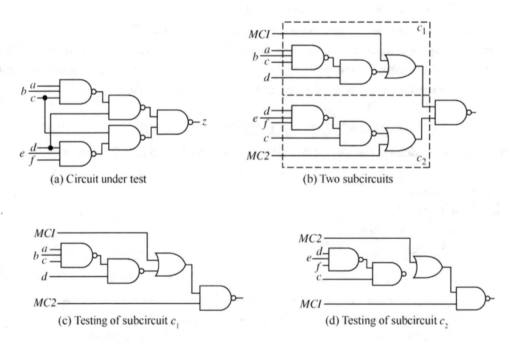

(a) Circuit under test   (b) Two subcircuits

(c) Testing of subcircuit c_1 (d) Testing of subcircuit c_2

FIGURE 4.2: Partitioning of a circuit for autonomous testing.

When MC1=0 and MC2=1, sub-circuit C_2 is disabled; sub-circuit C_1, shown in Figure 4.2c, can be tested by applying all possible input combinations at a, b, c, and d. Similarly, when MC1=1 and MC2=0, sub-circuit C_1 is disabled. In this mode, the circuit structure is as shown in Figure 4.2d and sub-circuit C_2 can be tested by applying all input combinations at c, d, e, and f.

When MC1=MC2=0, the circuit functions as the unmodified circuit except for the added gate delay. The advantage of the design method is that any fault in the circuit itself and the testing circuit is detectable.

A modified version of the autonomous testing, called the *verification testing*, has been proposed in Ref. [3]. This method is applicable to multi-output combinational circuits, provided each output depends only on a proper subset of the inputs. The verification test set for a circuit is derived from its *dependence matrix*. The dependence matrix of a circuit has m rows and n columns; each row

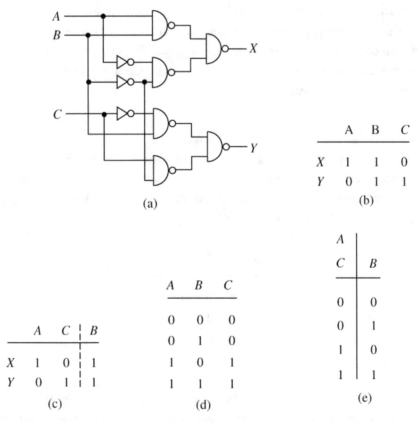

FIGURE 4.3: (a) Circuit under test. (b) Dependence matrix. (c) Partition-dependent matrix. (d) Verification test set. (e) Reduced verification test set.

represents one of the outputs, and each column one of the inputs. An entry $[(i, j): i=1...m, j=1, n]$ in the matrix is 1 if the output depends on input j, otherwise the entry is 0, m and n are the number of outputs and inputs respectively.

To illustrate, the dependence matrix for the circuit of Figure 4.3a is shown in Figure 4.3b. The dependence matrix is derived by tracing paths from outputs to inputs. A *partition-dependent matrix* is then formed by partitioning the columns of dependence matrix into a minimum number of sets, with each row of a set having at most one 1-entry; there may be a number of partition-dependent matrices corresponding to a dependent matrix. A partition-dependent matrix corresponding to the dependent matrix of Figure 4.3b is shown in Figure 4.3c.

A verification test set is obtained by assigning same values to all inputs belonging to the same partition of a partition-dependent matrix; any two inputs belonging to different partitions receive distinct values. Figure 4.3d shows the verification test set for Figure 4.3a. A reduced verification test set can be derived from a verification test set by removing all repetitions of identical columns. The reduced verification test set for the circuit of Figure 4.3a is shown in Figure 4.3e.

The verification testing is a useful technique for combinational circuits or circuits that can be transformed into combinational forms during testing (e.g., LSSD structure). However, the generation of the dependence matrix, which is the most important part of this test strategy, is a nontrivial task for circuits of VLSI complexity.

4.1.2 Pseudoexhaustive Pattern Generation

A combinational circuit with n inputs can be pseudoexhaustively tested with 2^n or fewer binary patterns if none of the outputs of the circuit is a function of more than w out of n inputs. For example, the following six test patterns can be used to pseudoexhaustively test a six-input circuit provided no output depends on more than two-input variables:

$$
\begin{matrix}
1 & 1 & 1 & 1 & 1 \\
1 & 0 & 0 & 0 & 0 \\
0 & 1 & 0 & 0 & 0 \\
0 & 0 & 1 & 0 & 0 \\
0 & 0 & 0 & 1 & 0 \\
0 & 0 & 0 & 0 & 1
\end{matrix}
$$

These test patterns can be generated by a non-primitive polynomial, for example, $x^6+x^4+x^3+x^2+x+1$. However, if an output of a multi-output circuit depends on more than two input variables, the derivation of minimal test patterns using a non-primitive polynomial may not be feasible.

In general, the pseudoexhaustive patterns needed to test an n-input and m-output combinational circuit are derived by using one of the following methods.

1. Syndrome driver counter;
2. Constant weight counter;
3. Linear feedback shift register /SR (LFSR/SR); and
4. LFSR/EX-OR gates (LFSR/EX-OR).

The *syndrome driver counter* method checks if p ($<n$) inputs of the circuit under test can share the same test values with the remaining ($n-p$) inputs [4]. Then the circuit can be exhaustively tested with 2^p inputs.

To illustrate, let us consider the circuit shown in Figure 4.4a. No output is a function of both c and d; thus, they can share the same input signal. Also, no output is a function of both a and b; thus, they can share the same input signal. Therefore, the circuit can be tested with 2^2 ($p=2$) combinations, as shown in Figure 4.4b. The syndrome driver can be a binary counter. The drawback of the method is that if the value of p is close to n, it requires as many test patterns as in exhaustive testing.

The *constant weight counter* method uses an x-out-of-y code for exhaustively testing an m-input circuit, where x is the maximum number of input variables on which any output of the circuit under test depends [5]. The value of y is chosen such that all 2^x input combinations are available from any m column of the code words.

For example, the circuit of Figure 4.5 has five inputs and three outputs, but none of the outputs depends on more than three input variables; thus, $m=4$ and $x=3$.

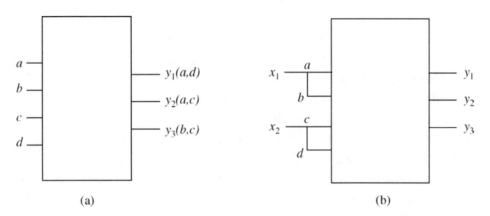

FIGURE 4.4: (a) Circuit under test. (b) Inputs a/b and c/d tied together.

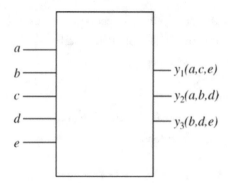

FIGURE 4.5: Circuit under test.

If y is chosen to be 6, the constant weight counter will generate the following 20 code words:

000111	100011
001011	100110
001101	100101
001110	101100
010011	101010
010101	101001
010110	110001
011100	110010
011010	110100
011001	111000

The selection of any five columns from the above code words will guarantee the exhaustive testing of the circuit associated with each output. In general, the constant weight test set is of minimal length. However, the complexity of the constant weight counter rapidly increases for higher x-out-of-y code.

An alternative approach for generating a pseudoexhaustive test set is to use a combination of an LFSR and an SR [6]. In an LFSR, the outputs of a selected number of stages are fed back to the input of the LFSR through an EX-OR network. An n-bit LFSR can be represented by an *irreducible* and *primitive* polynomial. If the polynomial is of degree n, then the LFSR will generate all possible 2^n-1 nonzero binary patterns in sequence; this sequence is termed the *maximal length sequence* of the LFSR.

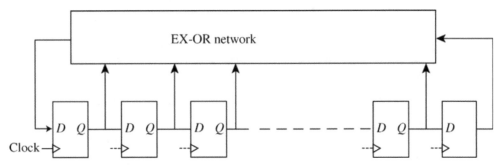

FIGURE 4.6: General representation of an LFSR.

Figure 4.6 shows the general representation of an LFSR based on the primitive polynomial:

$$P(x) = x^n + p_{n-1}x^{n-1} + \ldots + p_2x^2 + p_1x + p_0 \qquad (4.1)$$

The feedback connections needed to implement an LFSR can be derived directly from the chosen primitive polynomial.

To illustrate, let us consider the following polynomial of degree 4:

$$x^4 + x + 1.$$

This can be rewritten in the form of expression (4.1):

$$P(x) = 1 \cdot x^4 + 0 \cdot x^2 + 1 \cdot x + 1 \cdot x^0$$

Figure 4.7a and 4.7b shows the four-stage LFSR constructed by using this polynomial and the corresponding maximal length sequence, respectively:

Figure 4.8a and 4.8b shows the combination of a 3-bit LFSR and a 2-bit SR and the resulting output sequence. Two separate starting patterns are needed, one for LFSR and the other for SR:

The LFSR/SR combination of Figure 4.8 can be used to test any five-input circuit in which no output is a function of more than two input variables. This approach guarantees near minimal test patterns when the number of input variables on which any output of circuit under test depends is less than half the total number of input variables.

A variation of the LFSR/SR approach uses a network of EX-OR gates instead of an SR [7]. For example, the circuit of Figure 4.8a can be modified such that d is a linear sum of a and c, and e is the linear sum of b and d. The resulting circuit and the patterns generated by it are shown in Figure 4.9a and 4.9b, respectively. In general, the LFSR/EX-OR approach produces test patterns that are very close to the LFSR/SR approach.

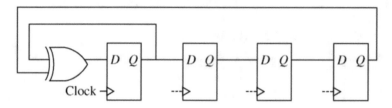

(a) A 4-bit LFSR

```
1 1 1 1
0 1 1 1
1 0 1 1
0 1 0 1
1 0 1 0
1 1 0 1
0 1 1 0
0 0 1 1
1 0 0 1
0 1 0 0
0 0 1 0
0 0 0 1
1 0 0 0
1 1 0 0
1 1 1 0
```

(b) Maximal length sequence

FIGURE 4.7: (a) A 4-bit LFSR. (b) Maximal length sequence.

An alternative technique based on LFSR/SR approach called *convolved LFSR/SRs* uses an n-stage SR and an LFSR of degree w to generate pseudoexhaustive test patterns for a circuit with n inputs and m outputs, with no output being a function of more than w input variables [8] .

The first step in test pattern generation using a convolved LFSR/SR to assign residues R_0 through R_i inputs 0 through i of the circuit under test. The residue of stage i is x^i mod $P(x)$ where $P(x)$ is the primitive polynomial used to implement the LFSR. To illustrate the computation of residues, let us consider a circuit with five inputs and five outputs as shown in Figure 4.10. The circuit has five inputs; hence, a five-stage convolved LFSR/SR is needed.

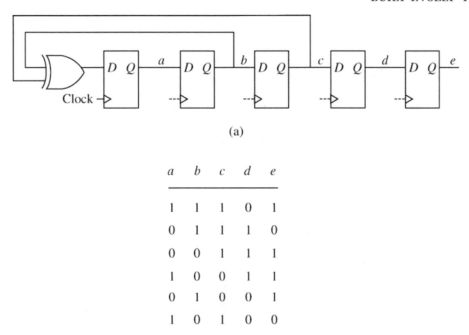

FIGURE 4.8: (a) LFSR/SR. (b) Output sequence.

Because $w=3$, the first three stages of the SR are used to implement a primitive polynomial of degree 3, e.g., x^3+x+1.

Figure 4.11 shows the resulting convolved LFS/SR. The residue of each stage is computed as x^i ($i=0$, …, 4) mod (x^3+x+1). For example, the residue of stage 3 is x^3 mod (x^3+x+1), i.e., $x+1$.

The next step in the test generation process is to assign residues to the inputs of the circuit under test so that for an output cone no assigned value is linearly dependent on already assigned values. For the circuit of Figure 4.10, inputs a, b, c, and d can be assigned residues 1, x, x^2, and $1+x$, respectively. However, input e cannot be assigned residue x^2+x because this will result in a residue set for output Z, namely, (R_1, R_2, R_4), which is linearly dependent. This can be avoided by assigning residue R_4 to input e.

If residue R_{i+j} is selected for assignment to input $i+1$ because residues $R_{i+1}, R_{i+2}, …, R_{i+j-1}$ cannot be assigned to this input because of linear dependence, then stage $i+1$ can be made to generate residue R_{i+j} by finding the linear sum of one or more previously assigned residues to stage $i+1$. For the circuit under consideration, stage 4 of the convolved LFSR/SR generates the desired residue x^2+x+1 by feeding the linear sum of residues R_2 and R_3 to stage 4 as shown in Figure 4.12.

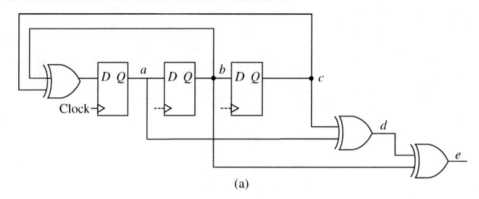

(a)

a	b	c	d	e
0	0	1	1	1
1	0	0	1	1
0	1	0	0	1
1	0	1	0	0
1	1	0	1	0
1	1	1	0	1
0	1	1	1	0

(b)

FIGURE 4.9: (a) LFSR/EX-OR. (b) Output patterns.

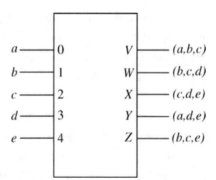

FIGURE 4.10: A circuit with n=5, m=5, and w=3.

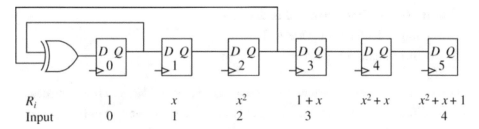

R_i	1	x	x^2	$1 + x$	$x^2 + x$	$x^2 + x + 1$
Input	0	1	2	3		4

FIGURE 4.11: Residue assignment for convolved LFSR/SR.

Assuming the initial seed for the LFSR to be 110 and that for the SR to be 011, the following pseudorandom patterns are generated by the convolved LFSR/SR:

$$
\begin{array}{cccccc}
1 & 1 & 0 & 0 & 1 & 1 \\
1 & 1 & 1 & 0 & 0 & 1 \\
0 & 1 & 1 & 1 & 1 & 0 \\
1 & 0 & 1 & 1 & 0 & 1 \\
0 & 1 & 0 & 1 & 0 & 0 \\
0 & 0 & 1 & 0 & 1 & 0 \\
1 & 0 & 0 & 1 & 1 & 1 \\
\end{array}
$$

Any five columns of these patterns constitute the minimal pseudoexhaustive test set for the output cones of the circuit of Figure 4.10.

4.1.3 Pseudorandom Pattern Generator
Pseudorandom patterns are sufficiently random in nature to replace truly random sequences. LFSRs are widely used for generating test patterns for combinational circuits because an LFSR is easy to implement. However, three related issues need to be considered in order to measure the effectiveness of pseudorandom testing [9]:

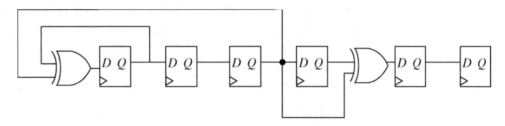

FIGURE 4.12: Convolved LFSR/SR.

1. Determination of the number of test patterns;
2. Evaluation of the fault coverage; and
3. Detection of random pattern-resistant faults.

The fault coverage can be evaluated by using exhaustive fault simulation. However, pseudorandom patterns needed to test a circuit are typically large; thus, fault simulation can be expensive. A relationship between a pseudorandom test sequence of length L and the expected fault coverage $E(c)$ is given in [10]:

$$E(c) = 1 - \sum_{k=1}^{2^n-1} \left(1 - \frac{L}{2^n}\right)^k \frac{h_k}{M}$$

where n is the number of circuit inputs, h_k (k=1, 2, 3, …, 2^n) is the number of faults in the circuit that can be detected by k input vectors, and M is the total number of faults in the circuit under test. It will be clear from the preceding expression that h_k needs to be known a priori to evaluate $E(c)$. The h_k for the two-input circuit of Figure 4.13 are (h_1, h_2, h_3, h_4)=(7, 0, 1, 0), as shown in Table 4.1. For complex circuits, the h_k are extremely difficult to derive, and they can only be approximated using probabilistic analysis.

A major problem associated with pseudorandom testing is that the number of patterns needed to detect a random pattern-resistant fault may be very large. For example, let us consider the stuck-at-1 fault α at the input of a 10-input AND gate shown in Figure 4.14.

It is clear that only test pattern $abcdefghij$=1111110111 can detect the fault. The probability of an LFSR generating this particular pattern is 2^{-10}. Thus, a huge number of pseudorandom patterns need to be applied to a circuit under test that may contain random pattern-resistant faults to guarantee a high fault coverage. The inherent weakness of pseudorandom patterns as far as the detection of random pattern-resistant faults is concerned arises because each bit in such a pattern has a probability of 0.5 of being either 0 or 1. If instead of generating patterns with uniform distribution of 0s and 1s, a biased distribution is used, there is a higher likelihood of finding test patterns for random pattern-resistant faults.

This is the principle of the *weighed test generation* technique proposed in Ref. [11]. An alternative way of generating pseudorandom patterns is to use a *cellular automation* (CA). A CA consists

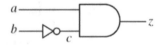

FIGURE 4.13: A two-input circuit.

TABLE 4.1: Fault detectability for the circuit in Figure 4.13

TEST	FAULT							
ab	a s-a-0	a s-a-1	b s-a-0	b s-a-1	c s-a-0	c s-a-1	z s-a-0	z s-a-1
00		×						×
01								×
10	×			×	×		×	
11			×			×		×
k	1	1	1	1	1	1	1	3
	$h_1 = 7$,		$h_2 = 0$,		$h_3 = 1$,		$h_4 = 0$	

of a number of identical cells interconnected spatially in a regular manner [12]. Each cell consists of a D flip-flop and combinational logic that generates the next state of the cell. Several rules may be used to compute the next state of a cell. For example, rules 90 and 160 are defined as:

$$\text{Rule 90:} \quad y_i(t+1) = y_{i-1}(t) \oplus y_{i+1}(t),$$
$$\text{Rule 160:} \quad y_i(t+1) = y_{i-1}(t) \oplus y_i(t) \oplus y_{i+1}(t),$$

where $y_i(t)$ denotes the state of cell i at time t. It has been shown in Ref. [13] that by combining these two rules, it is possible to generate a sequence of maximal length 2^s-1, where s is the number of cells in a CA.

To illustrate, let us consider the four-cell CA shown in Figure 4.15a. Assuming the initial state of the CA to be $abcd$=0101, the maximal length sequence for the CA is shown in Figure 4.15a.

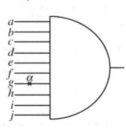

FIGURE 4.14: A 10-input AND gate with a stuck-at-fault.

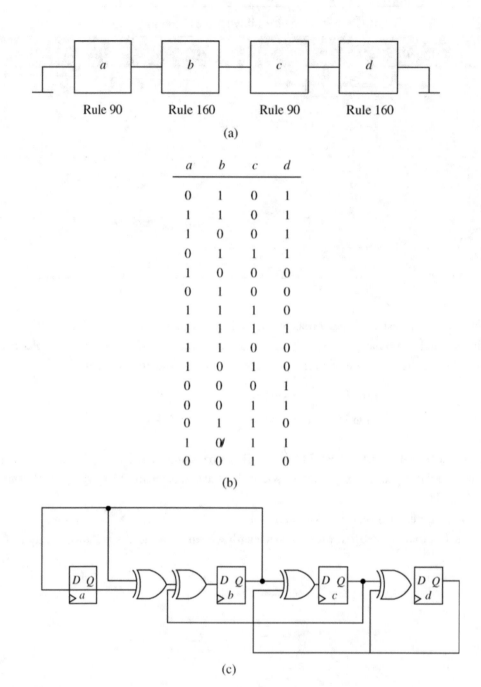

FIGURE 4.15: (a) A four-cell CA. (b) Maximal length sequence. (c) Implementation of the four-cell CA.

The implementation of the CA is shown in Figure 4.15c. Note that a 4-bit LFSR implementing a primitive polynomial of degree 3 will also generate a sequence of length 16. CAs based on rules 90 and 160 can generate all primitive and irreducible polynomials of a given degree [14]. Also, CAs do not require long feedback, which results in smaller delays and efficient layouts.

4.1.4 Deterministic Testing

Traditional test generation techniques may also be used to generate test patterns that can be applied to the circuit under test when it is in BIST mode. The test patterns and the corresponding output responses are normally stored in a read only memory (ROM). If the output responses of the circuit under test do not match the expected responses when the stored test patterns are applied, the presence of a fault(s) is assumed. Although, in principle, this is a satisfactory approach for fault detection, it is rarely used because of the high overhead associated with storing test patterns and their responses.

4.2   OUTPUT RESPONSE ANALYSIS

As stated earlier, BIST techniques usually combine a built-in binary pattern generator with circuitry for compressing the corresponding response data produced by the circuit under test. The compressed form of the response data is compared with a known fault-free response. Several compression techniques that can be used in a BIST environment have been proposed over the years; these include:

1. Transition count;
2. Syndrome checking; and
3. Signature analysis.

4.2.1 Transition Count

The transition count is defined as the total number of transitions of $1 \rightarrow 0$ and $0 \rightarrow 1$ in an output response sequence corresponding to a given input test sequence. For example, if an output response sequence $Z=10011010$, then the transition count $c(Z)=4$. Thus, instead of recording the entire output response sequence, only the transition count is recorded. The transition count is then compared with the expected one, and if they differ, the circuit under test is declared faulty [15].

Figure 4.16a shows the response sequences and the corresponding transition counts at various nodes of a circuit resulting from the application of a test sequence of length 4. Let us suppose there is a fault α s-a-0 in the circuit (Figure 4.16b). The presence of the fault changes the transition

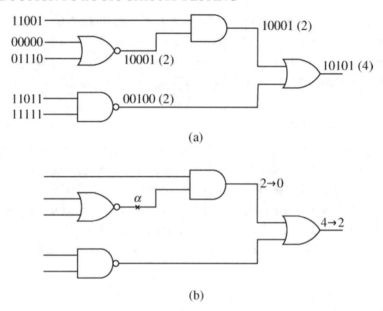

FIGURE 4.16: (a) Response to test sequence of length 4. (b) Changes in transition counts.

counts at certain nodes in the circuit (shown by *arrows*). As can be seen in the diagram, the transition count at the output node changes from 4 to 2, resulting in the detection of the fault α s-a-0.

The main advantage of transition counting is that it is not necessary to store the correct response sequence or the actual response sequence at any test point; only the transition counts are needed. Clearly, this results in the reduction of data storage requirements. However, this data compression may give rise to the *fault-masking errors*. This is because most transition counts correspond to more than one sequence; for example, the transition count 2 is generated by each of the following 6-bit sequences: 01110, 01100, 01000, 00110, 11011, and 10001. Hence, there is a possibility that a faulty sequence will produce the same transition count as the good sequence and therefore go undetected. However, as the sequence length increases, the hazard of fault masking diminishes.

4.2.2 Syndrome Checking

The syndrome of a Boolean function is defined as $S=K/2^n$, where K is the number of min-terms realized by the function and n is the number of input lines [16]. For example, the syndrome of a three-input AND gate is ⅛ and that of a two-input OR gate is ¾. Because the syndrome is a functional property, various realizations of the same function have the same syndrome.

The input–output syndrome relation of a circuit having various interconnected blocks depends on whether the inputs to the blocks are disjoint or conjoint, as well as on the gate in which the blocks terminate. For a circuit having two blocks with unshared inputs, if S_1 and S_2 denote the syndromes of the functions realized by the blocks 1 and 2, respectively, the input–output syndrome relation S for the circuit is:

TERMINATING GATE	SYNDROME RELATION S
OR	$S_1 + S_2 - S_1S_2$
AND	S_1S_2
EX-OR	$S_1 + S_2 - 2S_1S_2$
NAND	$1 - S_1S_2$
NOR	$1 - (S_1 + S_2 - S_1S_2)$

If blocks 1 and 2 have shared inputs and realize the functions F and G, respectively, then the following relations hold:

$$S(F + G) = S(F) + S(G) - S(FG),$$
$$S(FG) = S(F) + S(G) - S(\overline{FG}) - 1,$$
$$S(F \oplus G) = S(\overline{F}G) + S(F\overline{G}).$$

As an example, let us find the syndrome and the number of min-terms realized by the fan-out-free circuit of Figure 4.17. We have $S_1 = \frac{3}{4}$ and $S_2 = \frac{1}{4}$. Hence, $S_3 = 1 - S_1S_2 = 13/16$, and $K = S \cdot 2^n = 13$.

Table 4.2 lists the syndrome of the fault-free circuit of Figure 4.18, and the syndromes in the presence of fault α s-a-0 and the fault β s-a-1.

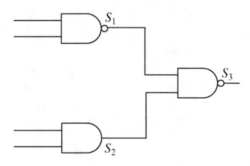

FIGURE 4.17: A fan-out-free circuit.

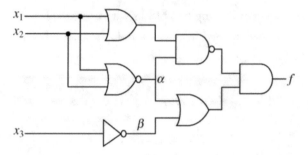

FIGURE 4.18: Circuit under test.

4.2.3   Signature Analysis

Signature analysis technique is pioneered by Hewlett-Packard Ltd. that detects errors in data streams caused by hardware faults [17]. It uses a data compaction technique to reduce long data streams into a unique code called the *signature*. Signatures can be created from the data streams by feeding the data into an *n*-bit LFSR. The feedback mechanism consists of EX-ORing selected taps

	TABLE 4.2: Fault-free and faulty syndromes		
	OUTPUT RESPONSE		
$X_1X_2X_3$	*FAULT-FREE*	α s-a-1	β s-a-0
000	1	1	1
001	1	1	1
010	1	0	0
011	0	0	0
100	1	0	0
101	0	0	0
110	1	0	0
111	0	0	0
Syndrome	5/8	2/8	2/8

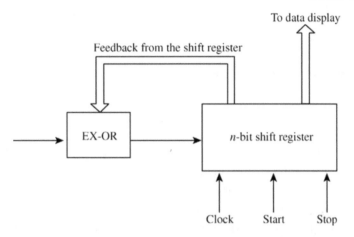

FIGURE 4.19: Signature analyzer circuit.

of the SR with the input serial data as shown in Figure 4.19. After the data stream has been clocked through, a residue of the serial data is left in the SR. This residue is unique to the data stream and represents its signature. Another data stream may differ by only 1 bit from the previous data stream, and yet its signature is radically different from the previous one. To form the signature of a data stream, the SR is first initialized to a known state and then shifted using the data stream; normally, the all-0 state is chosen as the initial state.

Figure 4.20a shows a simplified 4-bit signature generator. Assuming the content of the register is all 0, if a 1 is applied to the circuit, the EX-OR gate will have output 1. The next clock pulse will shift the gate output into the first stage of the register and 0s from the preceding stages into the second, third, and fourth stages, which leaves the register containing 1000, i.e., in state 8. From the state diagram of Figure 4.20b, the register contents or signatures can be identified for any data stream.

An n-stage signature generator can generate 2^n signatures. However, many input sequences can map into one signature. In general, if the length of an input sequence is m and the signature generator has n stages, then 2^m input sequences map into 2^n signatures. In other words, 2^{m-n} input sequences map into each signature. Only one out of 2^m possible input sequences is error-free and produces the correct signature. However, any one of the remaining $2^{m-n}-1$ sequences may also map into the correct signature. This mapping gives rise to *aliasing*; that is, the signature generated from the faulty output response of a circuit may be identical to the signature obtained from the fault-free response. In other words, the presence of a fault in the circuit is masked. The probability P that an input sequence has deteriorated into another having the same signature as itself is calculated on the assumption that any of the possible input sequences of a given length may be good or faulty:

$$P = \frac{2^{m-n} - 1}{2^m - 1}$$

For $m >> n$, the above expression reduces to:

$$P = \frac{1}{2^n}$$

Thus, the probability of aliasing will be low if a signature generator has many stages and hence is capable of generating a large number of signatures. For example, the 16-bit signature generator shown in Figure 4.21 can generate 66,636 signatures, and hence the probability that two input sequences will produce the same signature is 0.002%.

The error detection properties of the signature analysis technique are as follows.

First, the probability that two identical input sequences will produce the same signature is 1. Second, the probability that input sequences will produce the same signature if they differ precisely

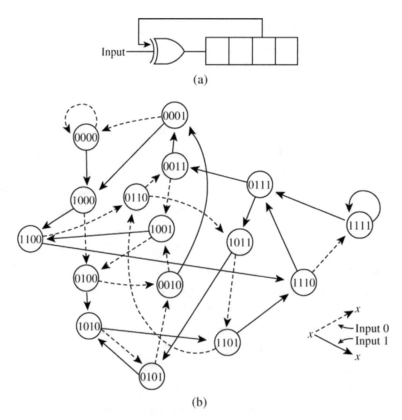

(a)

(b)

FIGURE 4.20: (a) A 4-bit signature generator. (b) State diagram of the signature generator.

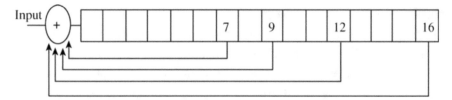

FIGURE 4.21: A 16-bit signature generator.

by 1 bit is 0. For example, let us consider two long input sequences, one differing from the other by only 1 bit. As the error bit gets shifted into the 16-bit register of Figure 4.21, it has four chances to change the signature register's input before it overflows the register (after 16 clock cycles) and disappears. The effect of the erroneous bit continues to propagate around the feedback network, changing the signature. For a single bit error, therefore, no other error bit comes along to cancel the feedback's effect, and so signature analysis is bound to catch the single bit errors. Single bit errors, incidentally, are typical of transient errors that occur in VLSI devices.

A signature, corresponding to the output sequence produced by a circuit under test, is usually created as discussed previously by feeding the sequence serially into the feedback line of an LFSR via an additional EX-OR gate. A signature can also be obtained by feeding a subset of the output sequence in parallel when a *multiple-input signature register* (MISR) is used. A k-bit MISR can compact an m ($>>k$)-bit output sequence in m/k cycles. Thus, an MISR can be considered as a *parallel signature analyzer*. Figure 4.22 shows an 8-bit MISR.

4.3   BIST ARCHITECTURES

Over the years, several BIST architectures have been proposed by researchers in industry and universities. We discuss some of these in this section.

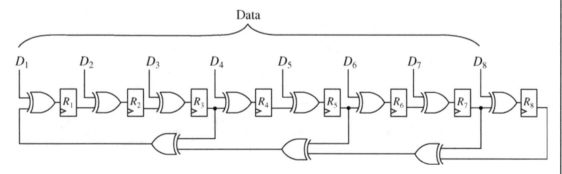

FIGURE 4.22: MISR.

4.3.1   Built-in Logic Block Observer

In built-in logic block observer (BILBO) structure, both scan-path and signature analysis techniques are employed [18]. It uses a multipurpose module, called a BILBO, that can be configured
to function as an input test pattern generator or an output signature analyzer. This is composed of
a row of flip-flops and some additional gates for shift and feedback operations. Figure 4.23a shows
the logic diagram of a BILBO. The two control inputs B_1 and B_2 are used to select one of the four
function modes:

1. *Mode 1. $B_1=0$, $B_2=1$.* All flip-flops are reset.
2. *Mode 2. $B_1=1$, $B_2=1$.* The BILBO behaves as a latch. The input data $x_1, \ldots, x_n$ can be simultaneously clocked into the flip-flops and can be read from the Q and Q outputs.
3. *Mode 3. $B_1=0$, $B_2=0$.* The BILBO acts as a serial SR. Data are serially clocked into the
   register through S_{in}, while the register contents can be simultaneously read at the parallel Q
   and Q outputs or clocked out through the serial output S_{out}.
4. *Mode 4. $B_1=1$, $B_2=0$.* The BILBO is converted into an MISR. In this mode, it may be used
   for performing parallel signature analysis or for generating pseudorandom sequences. The
   latter application is achieved by keeping $x_1, \ldots, x_n$ at fixed values.

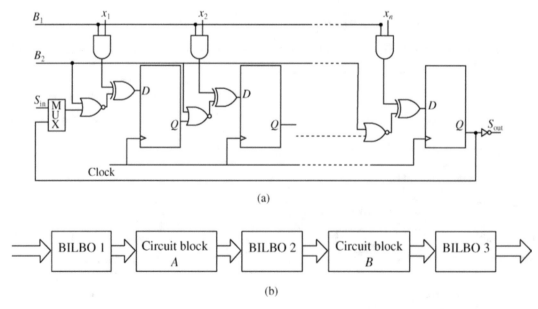

FIGURE 4.23: (a) Logic diagram of a BILBO. (b) BILBO-based BIST architecture.

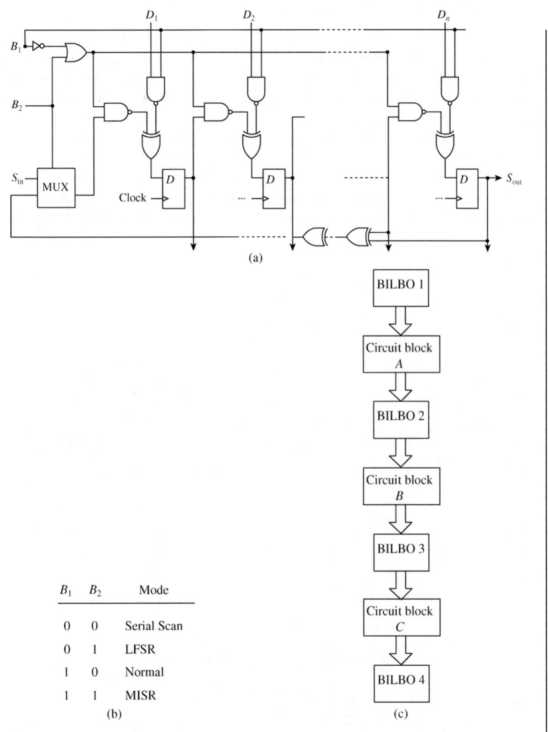

(a)

B_1	B_2	Mode
0	0	Serial Scan
0	1	LFSR
1	0	Normal
1	1	MISR

(b)

(c)

FIGURE 4.24: (a) Logic diagram of a modified BILBO. (b) Operating mode. (c) Simultaneous testing of pipeline structures.

Figure 4.23b shows the BILBO-based BIST architecture for two cascaded circuit blocks *A* and *B*. BILBO 1 in this structure is configured as a pseudorandom pattern generator, the outputs of which are applied as test inputs to circuit block A. BILBO 2 is configured as a parallel signature register and receives its inputs from circuit block A. Similarly, BILBOs 2 and 3 should be configured to act as a pseudorandom pattern generator and a signature register, respectively, to test circuit block B. It should be clear that circuit blocks A and B cannot be tested in parallel, because BILBO 2 has to be modified to change its role during the testing of these blocks.

A modified version of the conventional BILBO structure is shown in Figure 4.24a [19]. In addition to normal, serial scan, and MISR function, the modified BILBO can also function as an LFSR, thus generating pseudorandom patterns (Figure 4.24b). The modified BILBO can be used

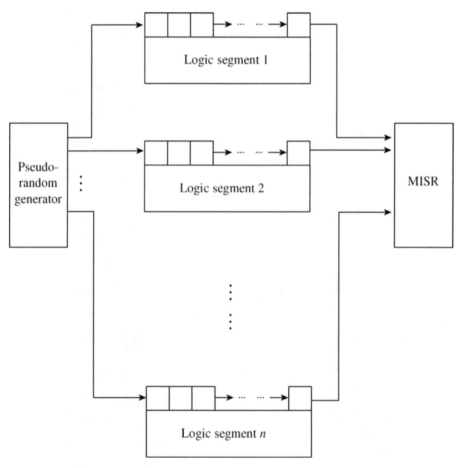

FIGURE 4.25: STUMPS configuration.

for simultaneous testing of pipeline structure. For example, in Figure 4.24c circuit, blocks A and C can be simultaneously tested by operating BILBOs 1 and 3 in the LFSR mode and BILBOs 2 and 4 in the MISR mode. Circuit block B can be tested individually by making BILBOs 2 and 3 operate in the LFSR and MISR modes, respectively.

4.3.2 Self-Testing Using an MISR and Parallel Shift Register Sequence Generator

Self-testing using an MISR and parallel shift register sequence generator (STUMPS) uses multiple serial scan paths that are fed by a pseudorandom number generator as shown in Figure 4.25 [20]. Each scan path corresponds to a segment of the circuit under test and is fed by a pseudorandom number generator. Because the scan paths may not be of same length, the pseudorandom generator is run till the largest scan path is loaded. Once data has been loaded into the scan paths, the system clock is activated. The test results are loaded into the scan paths and then shifted into the MISR.

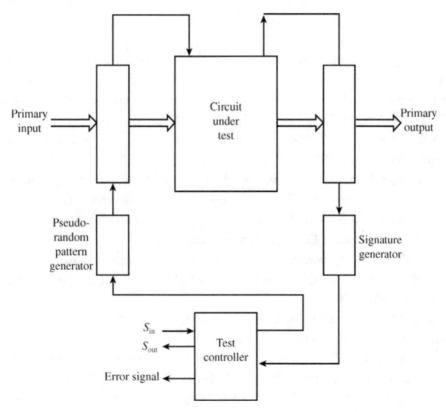

FIGURE 4.26: The LOCST configuration.

4.3.3 LSSD on-Chip Self-Test

LSSD on-chip self-test (LOCST) combines pseudorandom testing with LSSD-based circuit structure [20]. The inputs are applied via boundary scan cells; also, the outputs are obtained via boundary scan cells. The input and the output boundary scan cells together with the memory elements in the circuit under test form a scan path. Some of the memory elements at the beginning of the scan path are configured into an LFSR for generating pseudorandom numbers. Also, some memory elements at the end of the scan path are configured into another LFSR, which functions as a signature generator. Figure 4.26 shows the LOCST configuration.

The test process starts by serially loading the scan path consisting of input boundary scan cells, with pseudorandom patterns generated by the LFSR. These patterns are applied to the combinational part of the circuit under test, and the resulting output patterns are loaded in parallel into the scan path consisting of output boundary scan cells. These output bits are then shifted into the signature register. The resulting signature is then compared with the reference signature for verification purposes.

REFERENCES

[1] McCluskey, E. J., Logic *Design Principles: With Emphasis on Testable Semicustom Circuits*, Prentice Hall (1986).

[2] McCluskey, E. J. and S. B. Nesbet, "Design for autonomous test," *IEEE Trans. Circuits Syst.*, 10709 (November 1981). doi:10.1109/TCS.1981.1084930

[3] McCluskey, E. J., "Verification testing—pseudo-exhaustive test technique," *IEEE Trans. Comput.*, 541–6 (June 1984).

[4] Barzilai, Z., J. Savir, G. Markowsky, and M. G. Smith, "The weighted syndrome sums approach to VLSI testing," *IEEE Trans. Comput.*, 996–1000 (December 1981).

[5] Tang, D. L. and L. S. Woo, "Exhaustive pattern generation with constant weight vectors," *IEEE Trans. Comput.*, 1145–50 (December 1983).

[6] Barzilai, Z., D. Coopersmith, and A. Rosenberg, "Exhaustive bit pattern generation in discontiguous positions with applications to VLSI testing," *IEEE Trans. Comput.*, 190–4 (February 1983).

[7] Akers, S. B., "On the use of linear sums in exhaustive testing," *Proc. 15th Annual Symp. Fault-Tolerant Comput.*, 148–153 (June 1985).

[8] Srinivasan, R., S. K. Gupta, and M. A. Breuer, "Novel test pattern generators for pseudo-exhaustive testing," *Proc. Intl. Test Conf.*, 1041–50 (1993). doi:10.1109/TEST.1993.470594

[9] Agrawal, V. D., C. R. Kime, and K. K. Saluja, "A tutorial on built-in-self testing: Part 1. Principles," *IEEE Design Test Comput.*, 73–82 (March 1993). doi:10.1109/54.199807

[10] Wagner, K. D., C. K. Chin, and E. J. McCluskey, "Pseudorandom testing," *IEEE Trans. Comput.*, 332–43 (March 1987).

[11] Wunderlich, H. J., "Self test using equiprobable random patterns," *Proc. Intl. Symp. Fault-Tolerant Comput.*, 258–263 (1987).

[12] Wolfram, S., "Statistical mechanics of cellular automata," *Rev. Mod. Phys.* (July 1985).

[13] Hortensius, P. D., R. D. McLeod, and B. W. Podaima, "Cellular automata circuits for built-in-self-test," *IBM Jour. Res. Dev.*, 389405 (March–May 1990).

[14] Serra, M., T. Slater, J. C. Muzio, and D. M. Miller, "The analysis of one dimensional cellular automata and their aliasing properties," *IEEE Trans. CAD*, 767–78 (July 1990). doi:10.1109/43.55213

[15] Hayes, J. P., "Transition count testing of combinational logic networks," *IEEE Trans. Comput.*, 613–20 (June 1976).

[16] Savir, J., "Syndrome-testable design of combinational circuits," *IEEE Trans. Comput.*, 442–51 (June 1980).

[17] Hewlett-Packard Corp., *A Designer's Guide to Signature Analysis*, Application Note 222 (April 1977).

[18] Koenemann, B., J. Mucha, and G. Zwiehoff, "Built-in logic block observation techniques," *Proc. Intl. Test Conf.*, 37–41 (1979).

[19] Agrawal, V. D., C. R. Kime, and K. K. Saluja, "A tutorial on built-in-self testing: Part 2. Applications," *IEEE Design Test Comput.*, 69–77(June 1993). doi:10.1109/54.211530

[20] Bardell, P. H., W. H. McAnney, and J. Savir, *Built-In Test for VLSI Pseudorandom Techniques*, John Wiley and Sons (1987).

Author Biography

Parag K. Lala is the Cary and Lois Patterson Chair and Founding Chairman of Electrical Engineering at Texas A&M University–Texarkana; he was also the interim Chair of the Computer and Information Science Department at A&M–Texarkana for a year. Before his current position, he was the Thomas Mullins Chair Professor of Computer Engineering at the University of Arkansas at Fayetteville. He received his M.Sc.(Eng.) degree from the King's College, University of London, and his Ph.D. from The City University of London. His current research interests are in online testable logic, biologically inspired digital system design, fault-tolerant computing, and hardware-based molecular sequence matching. He has supervised more than 30 M.Sc. and Ph.D. theses and authored or coauthored over 135 papers.

He is the author of six books: *Fault-Tolerant and Fault-Testable Hardware Design* (Prentice-Hall, 1985), *Digital System Design Using PLDs* (Prentice-Hall, 1990), *Practical Digital Logic Design and Testing* (Prentice-Hall, 1996), *Digital Circuit Testing and Testability* (Academic Press, 1997), and *Self-Checking, Fault-Tolerant Digital Design* (Morgan-Kaufmann, 2001), and *Principles of Modern Digital Design* (John Wiley and Sons, 2007)

He was selected *Outstanding Educator* in 1994 by the Central North Carolina section of the IEEE. In 1998 he was awarded a D.Sc.(Eng.) degree in electrical engineering by the University of London for contributions to digital hardware design and testing. He was made a Fellow of the IEEE in 2001 for contribution to the development of self-checking logic and associated checker design. He is also a fellow of the IET (Institute of Science and Technology) in the United Kingdom.

Printed in the United States
by Baker & Taylor Publisher Services